Stories of Survival

STORIES OF SURVIVAL

Arkansas Farmers during the Great Depression

William D. Downs Jr.

THE UNIVERSITY OF ARKANSAS PRESS
FAYETTEVILLE   2015

Copyright © 2011 by William D. Downs Jr.
Reprinted by The University of Arkansas Press 2015.
All right reserved

No part of this publication may be reproduced, stored in a retrieval system, or transmitted in any form or by any means—electronic, mechanical, photocopy, recording, or any other—except for brief quotations in printed reviews, without the prior permission of the publisher.

ISBN: 978 1-55728-689-5
eISBN: 978-1-61075-575-7

19 18 17 16 15   1 2 3 4 5

Photograph, front cover: The frame of the author's grandparents' "Blue Farm" house built in the early 1920s in a remote area near Bexar (Fulton County), Arkansas.

This project was made possible by a grant from the Arkansas Humanities Council.

Library of Congress Cataloging-in-Publication Data

Stories of survival : Arkansas farmers during the Great Depression / [compiled by] William D. Downs, Jr.
   p. cm.
  ISBN 978-0-9824295-5-6 (alk. paper)
  1. Depressions—1929—Arkansas. 2. Farmers—Arkansas—Economic conditions—20th century. 3. Rural families—Arkansas—Economic conditions—20th century. 4. Farm life—Arkansas—History—20th century. 5. Arkansas—Rural conditions—History—20th century. 6. Agriculture—Social aspects—Arkansas—History—20th century. 7. Farmers—Akransas—Biography. 8. Arkansas—Biography. I. Downs, William David, 1932–
  F411.S86 2011
  976.7'052—dc22

2010050034

To my wife, Vera, for her constant encouragement

Contents

Preface and Acknowledgments	ix
I. Introduction to "Stories of Survival"	**1**
Why the Great Depression Was Late Coming to Arkansas	4
Cataclysmic Natural Disasters	6
•The 1927 Flood	6
•The Drought of 1930–1931	8
•Role of the Red Cross in Drought Relief	11
Plummeting Commodity Prices	15
•For Rural Arkansas, the Economic News Was All Bad	15
•Crisis in Bank Failures in Arkansas	16
Mounting Public Debt	18
•Arkansas in 1929: Forty-sixth in per Capita Income, First in per Capita Indebtedness	18
The Problem of Rural Arkansas's Self-Image	19
President Franklin D. Roosevelt Takes Office in 1933	20
The New Deal	20
State's Robinson Sold New Deal in Congress	21
FDR's First One Hundred Days	22
New Deal not without Its Critics	24
Establishment of the Dyess Colony	25
Sharecroppers and the Rise of the Southern Tenant Farmers Union	26
Arkansas after the Great Depression	28
II. Meet the Survivors	**29**
III. Stories of Survival: Arkansas Farmers during the Great Depression	**47**
Chapter 1: Earliest Memories of the Great Depression	49
Chapter 2: Sharecropping in Arkansas	61

Chapter 3: Income Sources for Buying Food, Seed, and Other Items	65
Chapter 4. Crops and Gardens and How They Were Worked	74
Chapter 5. Typical Chores around the Farm	86
Chapter 6. Preserving Milk, Making Butter and Buttermilk	93
Chapter 7. Hog Killing: A Special Time of Year	96
Chapter 8. Making Molasses, Peanut Butter, and Such	100
Chapter 9. Meeting and Taking Care of Hoboes	103
Chapter 10. Making Our Own Clothes, Going Barefooted	105
Chapter 11. How We Washed Our Clothes	113
Chapter 12: Family Bathing Routines	116
Chapter 13: Did We Know We Were Poor?	120
Chapter 14: Transportation—How We Got Around	130
Chapter 15: School Experiences	135
Chapter 16: How Did We Entertain Ourselves?	148
Chapter 17: Courtship and Marriage	161
Chapter 18: Medical Care	166
Chapter 19: Home Remedies	170
Chapter 20: To What Extent Did the Weather Affect Crops?	179
Chapter 21: Sources of News, Listening to Radio	186
Chapter 22: Christmas Memories	194
Chapter 23: Spiritual Life and Good Neighbors	206
Chapter 24: Serious Problems That Were Faced	215
Chapter 25: Worst Memories	221
Chapter 26: Effects of the Agricultural Adjustment Act	231
Chapter 27: Racism during the Depression Era	234
Chapter 28: Best Memories of the Depression	239
Chapter 29: Funniest Memories	248
Chapter 30: Other Memories	252
Chapter 31: Survival Stories Our Young People Need to Hear	278
IV. Epilogue	**291**
V. Appendix	**293**
Notes	297

Preface and Acknowledgments

Stories of Survival: Arkansas Farmers during the Great Depression is not intended to be a comprehensive history of the Great Depression. Instead, it is a quick overview of some of the Depression's main events in Arkansas and, more important, their impact on farmers.

In these interviews, readers will meet dozens of farmers who share their memories of the cultural, economic, and spiritual environments in which they lived. Invariably, when they were asked if they knew they were poor back in the 1930s, they replied with no trace of bitterness or regret, "Yes, we knew that we were poor. But so was everyone else." As noted by Nancy Weaver Williams of Snowball, Arkansas, "The Great Depression created a trend of resourcefulness. People had to learn how to live with what they had, and not to be overwhelmed by the sense of scarcity. Not only did people learn how to be resourceful," she added, "but they also developed a sense of community through this crisis."

The interviews were recorded exactly in the way they were spoken. In other words, no effort has been made to correct grammar. Without exception, they eloquently expressed the indomitable American spirit that responds to hardship with faith in God, courage, a sense of humor, and a love for their neighbors as they love themselves. Although I had intended to use at least one response from each survivor, I could not do so because in so many cases the responses were repetitious.

While I will leave it to others far more knowledgeable than I in the coming years to write a definitive comparison of the Great Depression with the current financial crisis America continues to face in 2010, I cannot completely ignore it. With this in mind, I talked with Dr. C. Fred Williams, director of the Center for Arkansas Studies at the Universty of Arkansas-Little Rock, who has identified what he considers to be some significant connections between the Depression of the 1930s and the situation today. Here was his reply:

- The Florida land boom of the 1920s and the subprime mortgages of 2004-2006 both created a "bubble" waiting to burst.
- Overexpansion in the construction industry (particularly in private housing) had a ripple effect throughout the "supplier chain" for appliances, carpets, wood products, etc., and collapsed in 1927 and in 2006.
- Production exceeded wages by a wide margin: manufacturing production increased 32 percent, while wages were up only 8 percent in

the 1920s. Similar numbers appeared in the first decade of the twenty-first century.

• Failure to regulate "big business" by withdrawing antitrust suits, by refusing to file new suits, and by appointing individuals to regulatory bodies who did not believe in governmental regulation.

• Allowing investment banking to use depositors' money for high-risk investments, a common practice before the Glass-Steagall Act was passed in 1934—prohibiting commercial banks from collaborating with full-service brokerage firms or participating in investment banking activities—and a return to high risk ventures (using someone else's money) after the G-S Act was repealed in 1996.

Ten years from now economists and historians will either marvel at how the Obama administration led America successfully through the Second Great Depression or they will shake their heads in disbelief that so many of our nation's leaders could have been so wrong.

Let's hope—and pray—for the best outcome. That's what the Arkansas farmers did in the 1930s.

Had it not been for the following people, this book would never have been written. First is my wife, Vera, for her insight, encouragement, proofreading, editing, and suggestions. Among those at the top of the list must be Jama Best, the senior project director of the Arkansas Humanities Council in Little Rock. Her steady encouragement, advice, patience, and guidance were invaluable. Also included here are John Coghlan, my Phoenix International publisher, for his interest in this project, and Debbie Upton, for her editing skills.

Also assured of my everlasting gratitude are the humanities scholars who reviewed the Introduction to the book and in doing so, greatly strengthened it in providing accurate review of the impact of the Great Depression on Arkansas farmers: Dr. Thomas A. DeBlack, associate professor of history at Arkansas Tech University; Tom W. Dillard, head of Special Collections, University of Arkansas Libraries, Fayetteville; Dr. Ben Johnson III, dean of Liberal and Performing Arts, Southern Arkansas University; Dr. Wendy Richter, Arkansas History Commission; Dr. Jeff Root, dean, School of Humanities, Ouachita Baptist University; Dr. Jeannie Whayne, Department of History, University of Arkansas, Fayetteville; Dr. C. Fred Williams, professor of history and director of the Center for Arkansas Studies, University of Arkansas, Little Rock; Dr. Trey Berry, deputy director for Museums of the Department of Arkansas Heritage, Dr. S. Ray Grenade, director of Library Services and professor of history, Ouachita Baptist University; and Allen Weatherly, executive director of the Arkansas Educational Telecommunications Network in Conway.

I also want to acknowledge the valuable feedback from my three sons, Bill III, Bob, and Ben, after they had plowed through the manuscript.

Ultimately, however, my deepest gratitude goes to the "Survivors" themselves, those who shared with me their in-depth memories of the Depression years. One of the most remarkable interviews was with Mrs. Union Henrietta Stoudamire, 103, a dignified African American resident of Pine Bluff, who loved a good conversation. Just after her 104th birthday in May 2010, she called me to ask when the book was coming out. When I told her it would probably be in the early part of 2011, I could hear her disappointment. A week later, I learned that Mrs. Stoudamire had died.

My hope, of course, is that her story and the many others that follow will keep these memories alive so that years from now, readers can reach across the widening chasm of time to again discover how these once-young and gallant men and women survived the Great Depression.

In addition, thanks to Betty Guthrie McCollum of Melbourne, Arkanasas, and Sue Shell Chrisco of Sage, Arkansas, for sharing excerpts from "Down Memory Lane"; Maggie King of Arkadelphia for sharing her interview with her grandfather, Jack Wood; and Nancy Williams Weaver of Snowball for sharing her research, "The Great Depression: A Terror or a Blessing?"

I. Introduction

My earliest memories of the Great Depression were in Calico Rock (Izard County), Arkansas, where I was born in 1932. Herbert Hoover was president, a gallon of milk cost 43 cents, a new car $540, average yearly income was $1,472, and gas was selling for 10 cents a gallon. It was also what some call the Depression's cruelest year. An indication of the Depression's enormous impact on our country is that it is frequently ranked along with the American Revolution, the Civil War, the second industrial revolution, and World War II as one of the formative events in our nation's history.

Among those rural memories are being flogged at the age of four by a big and very angry red rooster in our outhouse, puffing on an empty corncob pipe in front of Ethel and Marshall Floyd's hardware store, and my mother's invitation to President Franklin D. Roosevelt to come to Calico Rock to help me celebrate my fifth birthday—January 30—which was his birthday, too. On a "thank you" card I still have, the president sent his regrets.

Memories also include frequent visits with my grandparents, Charlie and Vollie Mae Hudson—"Pa" and "Ma"— at their dusty and remote little "Blue Farm" in the nearby Bexar (Fulton County) community. That's where I was kicked by a mule, traveled with my grandfather to deliver mail to a family that lived on a dirt floor, and climbed up on Pa's lap after Ma's supper each evening to hear stories about being chased by wild hogs or stalked by hungry wolves or the joys of hunting rabbits. The stories always began with the same conspiratorial whisper: *"Once upon a time, there were twooooo little boys, and theeeeese little boys were brothers"* or *"Sheep and a hog, walking through the pasture. Sheep says, 'Hog, can't you walk a little faster?' Hush, honey, hush! Bulldogs growlin'. Wolves in the woods—can't you hear them a-howlin'?"*

But it wasn't until almost sixty years later that I began to realize how tough things must have been for my grandparents. During a family reunion a few miles south of Bexar in 1989, I joined my mother, Ann Hudson Downs, and her two sisters, Billie Hudson Bagby and Cecile Hudson Metcalf—all three now gone and still missed—in returning to what little was left of the Blue Farm. Amazingly, most of the wooden frame was still standing with scraps of old newspapers dated in the 1930s still covering the cracks in the walls to protect against the winter winds. Yet not even once did I recall Pa and Ma complain about how bad things were during the Depression. The same can be said for the dozens of farm families I interviewed for this project.

My mother, Ann Hudson Downs, and her two sisters, Cecile Hudson Metcalf and Billie Hudson Bagby—all three now gone and still missed—in returning in 1989 to what little was left of my grandparents' Blue Farm

"Sure," they would say, "life was hard and we knew we were poor. But so was everybody else."

That return to Bexar led me to read David Kennedy's *Freedom from Fear: The American People in Depression and War: 1929–1945*. Hardest hit, he said, were American farmers. Only one house in ten had an indoor toilet, four out of five had no electricity. There were frequent illnesses, many of them medically untended, such as malaria, hookworm, and pellagra, which is caused by malnutrition. Just how bad things were was captured in the reaction of an unidentified English journalist in the 1930s. Although she had traveled widely through Europe and Africa, she had "never seen such terrible sights as I saw yesterday among the sharecroppers of Arkansas."

A similar reaction was expressed by then secretary of agriculture Henry

Wallace. After returning from a swing through the cotton states from Arkansas to the East Coast, Wallace said he had witnessed "a poverty so abject, that one-third of the farmers of the United States live under conditions which are so much worse than the peasantry of Europe, that the city people of the United States should be thoroughly ashamed."[1]

How, I wondered, could farm families have survived such conditions? Here are two of the many good answers I received. Recounting the hard times she remembered in the Augusta, Arkansas, area, Mrs. Vernon Massey wrote, "You are probably thinking nothing good could possibly have come out of such a time. But I think the strength of people and their faith in God carried them through this trying time in our country's history. Families were close, everyone had to work to survive. Our country weathered the crisis. Its leaders made mistakes, but many of their programs were productive, and with some changes are still being carried on today."[2]

Nancy Williams Weaver of Snowball, Arkansas, put it this way: "The strain of trying to feed and clothe an entire family when one out of every four United States citizens was unemployed is unimaginable to us today. The Great Depression affected not only those who lived through it but also all the generations that followed. People had to learn how to cope with the inadequacies of life, and this strengthened their resolve in making their children's lives better."[3]

So it was with these impressions in mind that I visited the State Archives at the Arkansas History Commission in Little Rock in late 2008. Fully expecting to see banner headlines on the microfilmed front pages of the *Arkansas Democrat* and the *Arkansas Gazette* (October 29, 1929) screaming in bold, black letters that economic disaster was upon us. After scanning that "Black Tuesday" issue, however, I found that I was wrong—not a word did I find about the Depression's impact on Arkansas. In New York City, however, the panic had begun and had even crossed the Atlantic Ocean:

"Havoc on stock market continues," cried the bold-face one-column page-one headline on October 29, 1929, *Arkansas Gazette*, followed by two stacked sub-headlines: Thousands who survived earlier slumps wiped out in latest crash, and HUGE LOSSES RESULT. A second page-one headline reported that "Panic seizes U.S. Tourists in Paris. "Panic scenes resembling those of the first days of the World War in August 1914, were being reenacted here today," the story began, "with hundreds of suddenly impoverished Americans, ruined by the series of Wall Street market crashes, seeking financial aid and steamship accommodations home."

Why the Great Depression Was Late in Coming to Arkansas

After reading these headlines and stories, I asked Russell Baker, the archivist, who happened to be standing behind me at the time, why bad news concerning the crash's effects on Arkansas did not appear on the front pages of our two statewide newspapers? He reminded me that not only the great majority of Arkansans—about eight out of ten— lived in rural areas in 1929, but that they had been suffering through their *own* depression during the 1920s—low cotton prices, the great flood of 1927, and a flurry of deadly tornadoes. Furthermore, since few if any Arkansans were investing in the stock market, the Wall Street crash in faraway New York City was the least of their worries. It was just as folk singer Woody Guthrie expressed it in his "Song of the South":

> Well, somebody told us Wall Street fell,
> But we were so poor that we couldn't tell.

Woody had it right. Hard times were nothing new to Arkansas farmers. Even though the 1920s had been a nightmare, particularly in the rural areas, the worst still lay ahead—the crippling droughts of 1930–1931 and the Great Depression itself. By 1933, the price of cotton had dropped to five cents a pound, far below the cost of production. In fact, no crop paid for itself, and merchants and businessmen who depended upon the health of the agricultural economy—which would have been most of them—were equally hard hit. Banks were closing their doors, leaving depositors stranded and without funds.

Living under almost primitive conditions, wrote historian Dr. Donald Holley, typical families had no electric lights, no indoor plumbing, no telephones, and few radios. Some Arkansans were even found living in caves.

Life had become so increasingly difficult for Ozark farmers, that B. C. Hall called the Great Depression "an epoch of continual bad times in which families lived on whippoorwill peas and poke salat; hoboes traveled the roads like tumble bugs; people guarded to the death their turnip patches with shotguns." When his father left home to work on the railroad, where he earned fifty cents a day cutting railroad ties, "we used to think that we were too poor to afford a father," Hall said. "There were nine of us kids. We slept in a one-room shot-gun shack and picked cotton from sun-up till dark for just enough money to buy lard and cornmeal," all of which he considered to be "wonderfully adventurous and exciting."[4]

How can such an experience be considered "wonderfully adventurous and exciting?"

Ironically, much of the answer may be found in the fact that since most Arkansans lived in rural areas, at least in one way they were better off than city dwellers—they could grow their own food. One of the survivors, Mary Leona Cartillar, who was interviewed on June 1, 2007, remembers it this way: "We raised everything that we ate. After we left where we was at, where Daddy was farming, we raised corn and tomatoes and okra. A many a time, I've gathered okra for breakfast. We raised food—that was all we raised. One of my sisters, she wouldn't eat vegetables. I loved them. We had chickens, and we had a milk cow."

In fact, the old saying "*Waste not, want not*" was a major theme during the Great Depression, said Nancy Williams Weaver. Most families grew what they could, canning "most of what they grew and gathered, such as wild huckleberries, blackberries and greens. They also had cows for milk, chickens for eggs and meat, and pigs for meat. too."[5]

During our interview, William Piercy of Bentonville recalled that his family's garden included potatoes, tomatoes, cabbage, pinto beans, butter beans, green beans, black-eyed peas, onions, sweet corn, sweet peas, popcorn, Swiss chard, spinach, and mustard greens. "We dried or canned or stored food for a full year out of that garden," he said. "My mother would can one thousand quarts and put them in the cellar. The smokehouse and cellar were locked during the Depression because people were starving and would steal food."[6]

As they increasingly felt the effects of the Depression, Arkansas farmers began to ask why was all this happening? *Who was to blame? How long was it going to last? And, most worrisome, how were they going to survive?* As Bible-belt Christians, some of the farmers even blamed themselves, assuming the Depression was God's punishment for their sins. For others, said Donald Holley, it was easier to blame bankers and businessmen. But drawing most of the blame was President Herbert Hoover. Interestingly, it was Hoover who had chosen the word "Depression" for what was going on in America because it sounded less frightening than "panic" or "crisis." Just how bad was the job market in Arkansas? According to William Manchester, it was bad enough that one Arkansas man walked nine hundred miles in 1932 in search of work.

To better understand the background and impact of the Great Depression on Arkansas, however, the shortest yet the most inclusive answer was provided by the historian Ben Johnson III, dean of liberal and performing arts and associate professor of history at Southern Arkansas University. "No single event brought the Depression to Arkansas," he said. "Instead, a series of interlocking crises swept over the state before the 1929 Wall Street crash: cataclysmic natural disasters, plummeting commodity prices, and mounting public debt."[7] Much of what follows is devoted to a closer look at these three factors.

Cataclysmic Natural Disasters

As if economic depression had not done enough damage to Arkansas in the aftermath of World War I, nature struck Arkansas from two extremes: First, the Great Flood of 1927, the most devastating natural disaster to hit the state since the New Madrid earthquakes of 1811–1812; and second, the Drought of 1930–1931, the worst in weather bureau records. During these abnormally hot and dry years, there was little rainfall, fields became dust bowls, and temperatures soared day after day to as high as 115 degrees. Ample evidence suggests that Arkansas might have suffered more flood and drought damage than any other state in the South or Southwest.

The 1927 Flood

Spawned by steady rains in the Mississippi River Valley during the winter and spring of 1927, the flood exploded through eight states—Illinois, Missouri, Kentucky, Mississippi, Louisiana, Oklahoma, Tennessee, and Arkansas, where it overflowed the banks of tributaries such as the Red River and the Arkansas River. In his riveting description of the surging flood in "Rising Tide," John Barry wrote, "It was as if the Mississippi was growing and swelling and rising in preparation, gathering itself for a mighty attack, sending out small floods as skirmishers to test man's strength. Those who knew the river always felt that it seemed a thing alive, with a will and a personality. In 1927 its will seemed intent on sweeping its valley clean of man."

Flood survivor Sam Huggins of Mississippi would later recall in his interview with Barry:

> The water rolled over and over itself, lifting trees, mules, roofs, dogs, cows and bodies, rolling forward, the water filthy, liquid mud, churning, spitting brown foam and froth. When the levee broke, the water just come whooshing, you could just see it coming, just see big waves of it coming. It was coming so fast till you just get excited, because you didn't have time to do nothing, nothing but knock a hole in your ceiling and try to get through if you could . . . It was rising so fast till peoples didn't get a chance to get nothing . . . People and dogs and everything like that on top of

houses. You'd see cows and hogs trying to get somewhere where people would rescue them . . . Cows just bellowing and swimming . . . A lot of those farmhouses didn't have no ceiling that would hold nobody."[8]

Flood waters from the Mississippi River spilled over into some areas of the Delta by as much as fifty miles, inundating more than 5 million acres in Arkansas, including 2,024,210 agricultural acres. By the time it had receded, the flood had caused 127 deaths; destroyed more than 55,000 homes; killed 25,000 horses, cows, mules, and hogs, which required twenty-five to thirty carloads of oil to burn their carcasses; amounting to more than $50 million in crop and property losses. In its wake, said the University of Arkansas historian Jeannie Whayne, the flood left stagnant pools of water and dead animals that aroused such major health concerns as typhoid, malaria, dysentery, and pellagra.

Finding shelter for the 143,000 people who had lost their homes was yet another concern. In her message to Secretary of War Dwight Davis, Arkansas senator Hattie Caraway wrote that "every available house and box car and tent at Helena and all of Phillips County is in use to house refugees from the overflowed section and still hundreds [are] unprovided for. Situation demands immediate action."[9]

McGehee, Ark., was inundated by the flood of 1927 *(Courtesy, Arkansas History Commission. All rights reserved)*

Such relief efforts for the 1927 flood victims were directed by U.S. secretary of commerce Herbert Hoover, which strengthened his successful campaign for the presidency in 1928. So far as flood damage was concerned, however, he drew criticism for insisting that repairing damaged roads and breached levees was the responsibility of the state rather than the federal government. Unfortunately, Arkansas had no resources to rebuild the infrastructure. As as result, responsibility for distributing food through landowners to displaced farm tenants was assigned to the Red Cross.

Despite the flood's devastation, the stock market crash, and several tornadoes that killed dozens of people and destroyed vast areas of farmland, by the spring of 1930, Arkansas farmers were confident enough to plant their crops and look forward to a bountiful harvest. But just days after the growing season began, the rains stopped. As they were soon to discover, Arkansas farmers were about to face the worst drought of the twentieth century.

The Drought of 1930-1931

Although twenty-three states across the Mississippi and Ohio river valleys and into the mid-Atlantic region were hit by the drought, eight southern states were particularly damaged, with Arkansas being the hardest hit of all. Crops burned up, the price of farmland plummeted, and many farmers lost their homes and their land. So severe were the effects of the drought that the plight of Arkansas farmers eventually attracted national attention.

No area of Arkansas escaped the effects of the drought, but some counties suffered more than others. Rainfall in Ashley County, for example, was 16 inches below normal. Helena on the Mississippi River and Fulton near Texarkana had gone for more than one hundred days without rain. During the summer of 1930, as noted by Tom Dillard, only 4.19 inches of rain had fallen on Arkansas, which was 35 percent of normal rainfall and the worst in the nation. As the severity of the drought increased, President Hoover was warned that farm families in the Lonoke area were facing starvation and that thousands of residents in the Texarkana area were desperate for food and winter clothing.

Recalling her childhood "in which hard times were traced by burrowing cracks in the dry earth," Sarah White Ragsdale described her memories of the drought:

> Whirling dust boils up from gravel roadbed, enveloping the trees and hollyhocks. Everything is the same monochromatic tan: people, trees, flowers and dogs. The choking, gasping heat of south Arkansas fills your throat.

> Why do I only remember the summertime? Later years and better years have winter, spring, and fall. But for me the summer and Depression are synonymous.[10]

So critical had life become that people began leaving their farms in search of food and work. Because ill-advised road-building programs had left the state so deeply in debt that it was unable to provide relief programs for its citizens, many farm families were loading their belongings on whatever transportation they had and heading for what was for most of them the empty promises of California.

In his classic 1939 novel, *The Grapes of Wrath,* John Steinbeck described the flight of families from the Dust Bowl:

> And then the dispossessed were drawn west from Kansas, Oklahoma, Texas, New Mexico, from Nevada and Arkansas, families, tribes, dusted out, tractored out car-loads, caravans, homeless and hungry, twenty thousand and fifty thousand and a hundred thousand and two hundred thousand. They streamed over the mountains, hungry and restless as ants, scurrying to find work to do—to live, to push, to pick, to cut—anything, any burden to bear, for food. The kids are hungry. We got no place to live. Like ants scurrying for work for food, and most of all for land.

In all 400,000 people left the Great Plains, victims of the combined action of severe drought and poor soil conservation practices.[11]

B. J. Kimbrough, writing in the weekly *Southern Standard* (Clark County, Arkansas), effectively captured the mood of the people in the following editorial:

> We are in the middle of a bad place caused by the long drought and low price[s] in 1930, and it seems that everybody, and even the government, is looking for a way out and, it seems that is a problem hard to solve, but the problem has got to be solved is an evident fact. I believe 90 percent of the farmers of Clark County will not be able to make a crop in 1931 unless they get help from some source. Now, here is the way the writer sees the situation. The biggest half of the people all over the county already owe the banks as much as they will take them for, so that puts them blank for another year and, it's an evident fact that people can't make a crop on the wind without food or stock feed.[12]

As expressed in their "Letters to Santa" in the *Southern Standard* on Christmas Day 1930, even the schoolchildren were aware of how tough times were:

> I am a little girl 8 years of age. I go to school. My teacher's name is Mrs. Gertrude Wetherington. I want a big red ball and a sleepy doll, some fruits, nuts and candy. Don't forget my little sister and brother, Ora and James. Old Santa, don't forget my teacher, Mrs. Gertrude Wetherington. Your little friend, Ruth Bradshaw, Gum Springs. P.S. And I want a box of colors.

> Dear Santa Claus: I am a little boy six years old. I am in the first grade, and I go to school every day. I love my teacher. Santa, I am going to tell you what I want for Christmas. I want an air-gun and a big rubber ball and all kinds of fruits and nuts. Santa, I won't ask for any thing more. Your little friend, Wayne Dee Pollard, Arkadelphia.[13]

With no relief being provided by the state, Arkansans looked to President Hoover for assistance. Unfortunately, he remained stubbornly opposed to providing direct relief to Arkansas and other drought-stricken states, saying that such action was not the responsibility of the federal government. To his credit, however, as pointed out by the historian Roger Lambert, the president did offer some suggestions on drought relief, including the encouragement of private, state, and federal loans, expansion of employment by restarting road-building programs, asking railroads to lower shipping rates for food and so forth, and requesting the Red Cross to provide relief to those who needed it most.

For whatever the reasons might have been, however, the first three of the four proposals were defeated by the governors, which left the Red Cross to be the only source of hope for those suffering through the worst drought on record.

In early 1931, Hoover surprised congressional leaders when he introduced a $45 million relief bill for loans to drought-stricken farmers. Incredibly, however, there was only one hitch: *The money could be used only to buy seed, fertilizer, and feed for livestock . . . but not for food for his family!* Infuriated, Senator Joseph T. Robinson of Arkansas, who had been leading the fight in Congress for the drought sufferers, lambasted the bill. His voice dripping with sarcasm, he said, "It is all right [for the President] to put a mule on the dole, but it is condemned, I see, to put a man on parity with a mule."

Eventually, however, Hoover accepted a Robinson-sponsored bill providing food loans. But since they couldn't afford to mortgage their crops for the following year, most farmers did not qualify for federal assistance. "The only immediate source of help was private charity,"[14] which meant once again that the Red Cross was virtually alone in providing relief.

Role of the Red Cross in Drought Relief

In some counties, practically everyone was living on Red Cross rations, said Donald Holley. In Chicot County, almost 22,000 people out of a total population of 22,646 received Red Cross aid each day. Rations consisted mostly of cornmeal, beans, and flour—at an average cost of sixty cents per person per week. Despite the enormity of the task the national Red Cross office sent only 40 relief experts into Arkansas, which left most of the work to be done by more than 6,000 volunteers.

Noting the deplorable conditions in which the Red Cross workers provided aid, Gail S. Murray described life in the plantation counties of eastern Arkansas:

> There the Negro and white sharecroppers were already living a hand-to-mouth existence *before* the drought and the depression. They worked for fifty of seventy-five cents a day and often bought all supplies at the "plantation commissary" where prices were high. They borrowed money on their crops from the plantation owners, often at interest rates as high as 25 percent. At the end of the year, the balanced statements almost always showed the sharecroppers in the red. When the effects of the Depression began to be felt in 1930, the cropper discovered that the plantation owner would no longer "tide him over" through the winter. Neither could a sharecropper get credit at local banks or qualify for a government loan. Even the Red Cross sometimes turned down his application for relief, considering him the responsibility of his plantation owner.[15]

Further documenting "the stark images" in eastern Arkansas tenant cabins, Ben Johnson wrote that a Red Cross worker had found the tenants to be "barefoot and without decent clothes, no meal, no flour in the bin, ragged children crying from hunger . . . nothing but hunger and misery . . . far worse than the Mississippi flood."[16] In the midst of the drought, many families "were out of luck. They had to sell or give animals away because they could not afford to feed that many mouths. One father paid his children a dime every evening to skip dinner—then in the morning, the children gave back the dime to be able to eat breakfast."[17] What it all came down to, observed Mrs.

Vernon Massey, "People made do with what they had." And because cash money was scarce, families accepted a variety of garden produce in place of currency, such as home-cured meat, chickens, watermelons, fruit, eggs, butter and milk, turnips and turnip greens. Even a cow.

> Although I had churned and cared for the milk and butter, I had never gotten close enough to a cow to milk her. It was hard to stay in business without much cash, but many people did. Food was very scarce in some instances; a friend said since they lived on a farm, they had plenty to eat and share with neighbors, but had difficulty getting clothes, school books, coal oil, and supplies they did not raise. Those with no garden didn't fare as well. An acquaintance told me they lived for a week on turnips; and she emphasized, only turnips![18]

So popular had turnips become, that a Little River County resident suggested that the old camp meeting song, *"Will there be any stars in my crown?"* be changed to *"Will there be any turnips in my crown?"* Even during the winter of 1930–31, when a third of the entire population of Arkansas was literally facing starvation, farmers could joke about the recipe for a turnip sandwich: *"three slices of turnips and put one in the middle."*[19]

The severity of the drought was captured in a 1931 issue of the *Literary Digest*:

> Arkansas, there she stands—withering. Like a burgeoning tree caught by the blight, a flower scorched by fire. Land of dusty desolation, sucked dry by the drought, rasped by the sharp edges of a cruel economic fate, her citizens destitute, hungry and helpless. By tens of thousands, they have turned to the Red Cross for food to keep them alive, waiting in line for their ten-cents-a-day ration, a tired and bewildered people clinging forlornly to the only thing they have left—hope.
>
> Arkansas was a land "where men, women, and children have lived for days on turnips, herbs, and roots, and on nuts gathered in the woods. Thousands of undernourished children are unable to 'sign up' with the Red Cross, each giving three references, in order to obtain the $2 weekly food allowance granted each family plus fifty cents for each child, the maximum being $4.50 a week. Mules are dying of starvation, and the survivors are so weak they are unfit for work. Meat is such a luxury that a rabbit, shot in the swamps by a lucky hunter, is called a 'Hoover Hog.'"[20]

Although the drought crisis had not aroused any organized resistance among Arkansas farmers, in August 1930, dozens of people seeking foodstuff raided stores in Conway, Fort Smith, and Pine Bluff. But it was not until early the next year that their anger, frustration and fear led to an event that brought nationwide attention to drought victims in Arkansas—"The England Food Riot," so named by the *New York Times* for an event that never really happened.

The England Food Riot

The trouble began on January 3, 1931, when the severity of the Depression and the drought finally combined to bring together an orderly but hungry and very determined crowd of farmers to gather in downtown England, Arkansas, where they threatened to loot grocery stores unless they were given food. The farmers had good reason to be both angry and frightened—their crops had been destroyed by the drought, cotton prices were continuing to fall, the local bank had closed and in the process had wiped out their savings, and to make matters even worse, the Red Cross had run out of "blanks" to issue free food.

What had come dangerously close to erupting into an actual "riot," however, was averted when H. C. Coney, a local tenant farmer who had been following congressional debates on relief efforts in Arkansas, was informed by a widow that her children had not eaten in two days. After saying, "Lady, you wait here. I'm going to get some food,"[21] he jumped into his truck and headed for the Red Cross office where he found about three hundred desperate farmers on that Saturday morning who were demanding that the lack of "blanks" was no reason for denying them food.

Trying to calm them down, Red Cross officials promised to get a new supply of blanks on the following Monday. In what England resident Charles Walls called "an orderly but good-natured" response,[22] the farmers said that because their wives and children were hungry, they couldn't wait that long. It didn't help that some of them had misinterpreted the Red Cross's request to mean that there would be no more aid at all, which was not true.

Fearing that a riot was imminent, the England business leaders contacted the Red Cross office in Kansas City, where they secured an agreement to reimburse the merchants for the food they would provide locally. After more than 150 loaves of bread were distributed to the needy families, the crowd began to disperse peacefully. But, said Gail Murray, their protests had brought nationwide attention to the realities of the drought and Depression and averted further trouble. The event even drew the attention of the humorist Will Rogers, who said the farmers had "hit the heart of America." He later did benefit performances in the state to raise money for the drought victims.

An interesting sidelight on how Mr. Coney became the natural spokesperson for the farmers was related by Lement Harris. In recording the interview, Harris said he had taken great care to capture "the flavor and tang to [Mr. Coney's] vernacular."

> Well, here's how it happened. We all got pretty low on food out here, and some was a-starvin'. Mebbe I was a little better fixed than most, 'cause we still had some food left. But when a woman comes over to me a-cryin' and tells me her kids hain't et nothin' fer two days, and grabs me and says, "Coney, what are we a-goin' to do?" Then I cranks up my truck you see settin' over yonder, and takes my wife and rolls over to Bell's place. Bell's the feller them Red Cross guys picks to run the relief, but he never give out nothin'. He always tells 'em that he hain't got no blanks and they gotta wait.
>
> Well, I rolls over to Bell's place and finds a crowd of hungry men and Bell still a-sayin' that he haint got no blanks. So I hollers out, "All you that hain't yaller, climb on my truck. We're all a-goin' into England to get some grub." They all loads onto her—forty-seven clum on, and let me tell you there warn't a one among 'em that had a gun of any sort. "Now then, when we gets to town," I says, "we'll ask for food quiet-like, and if they don't give it to us, we'll take it, also quiet-like." I and another did all the talkin'. The gang jest kept silent. They was a right pathetic sight.

The "riot" ended quickly, Coney said, when they doled out the feed and we all rolled back here without nobody gettin' hurt.[23]

Although some said the demonstration was staged and an Indiana congressman even went so far as to suggest it had been inspired by communist subversion, England's mayor said he "could not recall sighting any bolsheviks."[24] The incident amounted to nothing more than farmers trying to save their families. Even though the situation ended peaceably, that didn't prevent the national press from portraying it as the "England Food Riot." But it did have some positive results.

"The confrontation also led to as national debate on relief in general and *drought* relief in particular," said Tom Dillard. "President Herbert Hoover and Congress were forced to face the reality of the suffering, even if they did nothing immediately. As observed admiringly by Will Rogers, 'Paul Revere just woke up Concord. These birds woke up America.'"[25]

Elsewhere, the challenge of feeding hungry Arkansans was being met at a staggering rate. Between January and March in 1931, for example, the Arkansas Red Cross reported that 519,000 people received food allowances and that others were being served through the hot-lunch programs in the public schools.

> The goal of the Red Cross was to provide needy families with $2.00 per week food allowance with an additional fifty cents available for each child. The maximum any family could receive was $4.50 per week. However, the actual average ration in Arkansas was closer to $1.20 per person per month. This meant that a family of five would receive a $6.00 monthly allowance that usually consisted of 25 pounds of meal, 48 pounds of flour, 25 pounds of pinto beans, 2 gallons of molasses, 10 pounds of lard, 1 1/2 bushels of potatoes, 10 pounds of salt, one package of soda, and one can of baking powder. Obviously absent, she added, were such staples as meat, milk, fruit and sugar.[26]

As the drought-ending spring rains began to fall in 1931, Arkansas began to recover. But for some the rains had come too late. Since the cotton crops had been destroyed by the drought, buying on credit disappeared, which meant that Delta sharecroppers could no longer buy on credit from local merchants and plantation stores. And when forty-three banks closed on November 17, even affluent farmers were suddenly without credit.

Plummeting Commodity Prices
For Rural Arkansas, the Economic News Was All Bad

In addition to the the stock market crash and natural disasters such as the 1927 flood and killer tornadoes, soon to be followed by the record-breaking droughts in 1931, Arkansas farmers, particularly those who raised cotton, had been struggling through most of the 1920s with low agricultural prices. Although cotton was selling for seventeen cents per pound before the crash, by 1930 it was down to nine cents, then six cents in 1931. By 1933, the cotton market dropped to five cents, the lowest price in fifty years.

As income from farm production had fallen 60 percent—from $194,350,000 in 1929 to $78,451,000 in 1933, tax delinquencies increased to record levels. By 1933, almost a third of the total land area—10,151,147 acres—

was either delinquent or forfeited, sometimes at incredibly low amounts. "I have made every effort possible," said one farmer, "to save my 80-acre farm which will be certified to the state unless I can raise $28.34 for taxes."[27]

Crisis in Bank Failures in Arkansas

By 1932, farm incomes had plummeted "and nearly 40 percent of the labor force was unemployed. Bank deposits dwindled, and a record number of banks closed across the state as anxious depositors rushed to withdraw what they could." The numbers are staggering: Before the stock market crash in 1929, Arkansas had 347 banks. Four years later, only 202 were still in business. It was like the domino effect: When many of the merchants who had suffered significant losses could no longer sell on credit to planters, for example, planters were suddenly unable to 'furnish' their tenants and sharecroppers, who were suddenly left jobless and homeless with nowhere else to turn."[28]

Recalling how many people had lost large amounts of money when the Bank of Augusta closed, Mrs. Vernon Massey recalled that

> the stockholders were heavy losers; and although most depositors did eventually receive some of their money back, the people who suffered the most and the greatest losses were farmers who had borrowed money from the bank, giving mortgages on their homes, farmlands, stock, and equipment. The sad part about the whole thing was that times had appeared to be booming, the stock market was going up, when suddenly the bottom fell out! It was the worst of times, but some good came of it—today we have the FDIC, stricter rules relating to loans by banks, more frequent auditing, and more control of interest.

"I don't remember hearing anyone casting any reflections on the officers or employers of the bank, or the employees, as to their honesty," she added. "It wasn't their fault when cotton, the main crop of the farmers, fell from thirty cents a pound to less than five cents."[29]

The following column, which was published in the *Yellville Mountain Echo*, September 3, 1931, in the midst of the Depression, provides insights into the local impact of bank closures.

Citizens Bank closes its doors

On December 17th, 1930, the Citizens Bank in [Yellville] closed its doors, and the bank was turned over to the banking department.

A statement filed in the county clerk's office by the deputy state banking commissioner at that time showed individual deposits of $160,030.54, and savings deposits of $61,990.93. Included in this amount was between $27,000 and $28,000 of public funds, which belonged to the taxpayers of the county.

Everything moved along nicely, and the people had begun to believe the new institution was going to be a success. The public was informed that the thousands of dollars worth of notes of the defunct bank that had been put up as collateral in various banks had been redeemed with the exception of a few thousand dollars, on which plenty of time for their redemption has been given . . .

Imagine the consternation of our people when shortly after the bank had opened at 9:00 o'clock Tuesday morning, they were again closed, the blinds pulled, and the word given out that the directors and stockholders were holding a meeting in the directors' room of the bank.

The public waited patiently on the outside, wondering what the result might be. The result was not long delayed for about 10 o'clock the following notice was posted on the door: "This bank in hands of State Banking Commissioner." Thus, for the second time within twelve months the depositors found their money in a defunct bank—the taxpayers realized that the public funds of the county had been taken from them . . . .

[The closing has] almost made paupers of some of our best citizens, has depleted our public funds—school, road and general, has paralyzed business, and caused our people to lose confidence in banks."[30]

Although aware of how difficult life had become for its readers, the *Echo* also pleaded for payment of long-overdue subscriptions:

On account of the financial depression caused by the drouth [*sic*] of the past year, together with the bank failure just before the first of the year, the *Echo* made no effort to collect local subscription due on the first of the year, and we have been carrying the majority of our county subscribers, many of them for the past two years. While we have been glad to do this, it has been hard for us to do, as we, like all others, have suffered from the scarcity of money and business.

Since the county is blessed with bountiful crops this year, we feel sure that many of our subscribers will be able to pay up in full in the near future. Where it can be done, we would prefer that payment be made in cash, but if this cannot be arranged, we can use several ricks of wood and can also use some corn or wheat, if some of our subscribers have a surplus on hand.

Even in the midst of the crisis, however, there was occasionally some room for humor, as in the following story that appeared in the same issue:

> Over at Jonesboro, Joe Jeffers, a Baptist evangelist, who is conducting a revival meeting in that city, has raised so much hell among his own members that fist fights right in the church became so prevalent that the state militia had to be called out to maintain order while Jeffers poured "hot shots" into their ranks. We never knew before that the command given to the disciples, to "go preach Christ and Him crucified," produced that kind of a result among men but everything nowadays is changing so rapidly that the results at Jonesboro may be in keeping with the times, and is all right. It is to be hoped that those who survive the conflict will be better, if not wiser, men.[31]

Mounting Public Debt
Arkansas in 1929: Forty-sixth in per Capita Income, First in per Capita Indebtedness

As noted earlier, Arkansas farmers had been suffering their own depression for years before the stock market crash. In a state where cotton was king, cotton prices had declined and the Great Flood of 1927 had wreaked economic havoc. Governor John Martineau, a Democratic progressive, had won the gubernatorial election on his "better roads and better schools" slogan.

Even though he was aware that county road improvement districts were building more roads than they could pay for, Governor Martineau also knew that he had to take some sort of action in order for Arkansas to qualify for federal highway funding. With that in mind, he signed a measure that not only obligated the state to take over from local authorities the task of building highways but also included new revenue for road improvements and authorized the state to build even more highways. As a result, said historian C. Fred Williams of the University of Arkansas at Little Rock, Arkansas was suddenly burdened by $160 million in additional debt even though its entire annual budget was just over $14 million.

When Harvey Parnell became governor of Arkansas in 1929, the state highway construction program Parnell inherited from Governor Martineau was considered the national model for modern highway administration. Although Governor Parnell had secured the funds to expand the road program, said John Hume, Arkansas and other rural states were still trying to recover from the economic slump they had experienced since the end of World War I.

Furthermore, just three years after the Great Flood, Arkansas was struck by a second natural disaster—the record-breaking droughts of 1930–1931. Unfortunately, because debt payments arising from the questionable road-building decision were reducing available revenue by about 75 percent, when the drought struck Arkansas, the state had no budget reserves to meet the crisis.

Adding to the financial misery were the substantial numbers of Arkansas taxpayers who were refusing to pay their road, drainage, real estate, and property taxes. As the farming sector of the economy began to fail, said historian Jeannie Whayne, so did many of the state's merchants, hardware stores, and farm equipment suppliers that the farmers had supported. The result: calamitous declines in tax revenue led to a faltering economy so severe that the state was unable to provide relief that was so desperately needed.

Ranked forty-sixth in the country in per capita income in 1929 and first in per capita indebtedness, the state could no longer pay its bills. To illustrate just how bad things were, the General Revenue Fund balance on June 30, 1932, was $4.80. As a result, in 1933 Arkansas was the only one of the forty-eight states to default on its bonded indebtedness—$160 million. What this meant was that out of its annual $14 million budget, the state spent $13 million on debt service for roads. Defaulting on its indebtedness was seen in the nation's financial circles as a major embarrassment for Arkansas, one from which it is still suffering (see "The Problem of Rural Arkansas's Self-Image" that follows). During the next two years, state government functioned mostly on federal money.

Although the Futrell administration is said to have done little to combat the effects of the Depression, it did offer relief through a sales tax and other fund-raising measures such as dog racing and horse racing, including paramutual betting. Also, the federal government stepped in to provide much-needed programs such as the Works Progress Administration (WPA) and the Civilian Conservation Corps (CCC). But so far as road-building plans were concerned, Williams said they were put on hold for the next sixteen years.

The Problem of Rural Arkansas's Self-Image

Saddled as it was in the 1930s with a reputation for backwardness and poverty, "Arkansas desperately needed to feel good about itself," said Donald Holley. The nation was painfully aware that the Red Cross had fed a half million starving Arkansans during the drought of 1930–1931, that the state had defaulted on its bonded debt, and that the oppression of Arkansas sharecroppers and the conditions in which they lived were the worst in the United States.

Almost surely without intending to harm the state, humorists such as Bob Burns with his imaginary stories of his rural kinfolks, and the popular *Lum 'n' Abner* programs with Chester Lauck and Norris Goff had done much to create Arkansas' backwoods image.[32] Not to be overlooked was the *Baltimore Sun's* H. L. Mencken, who once described Arkansas as "perhaps the most shiftless and backward state in the whole galaxy. Only Mississippi offers it serious rivalry for last place in all American tables of statistics." Unfortunately, it's a problem that persists to this day.

President Franklin D. Roosevelt Takes Office in 1933

On the edge of bankruptcy toward the end of the Hoover administration, the mood not only in Arkansas but also in the nation was one of fear and panic when Franklin D. Roosevelt, the first Democratic president in sixteen years, was inaugurated on March 4, 1933. The historian Arthur Schlesinger captured this mood when he wrote that it "was now a matter of seeing whether a representative democracy could conquer economic collapse. It was a matter of staving off violence, even—some thought —revolution."[33]

As of this writing, millions of people still recall President Roosevelt's famous radio "Fireside Chats." Even on the day he was sworn into office, the country was in the midst of yet another bank panic. It seemed to serve as the backdrop of the new president's unforgettable reassurance to a frightened America:

> This great Nation will endure as it has endured, will revive and will prosper. So, first of all, let me assert my firm belief that the only thing we have to fear is fear itself—nameless, unreasoning, unjustified terror which paralyzes needed efforts to convert retreat into advance. In every dark hour of our national life a leadership of frankness and vigor has met with that understanding and support of the people themselves which is essential to victory. I am convinced that you will again give that support to leadership in these critical days. ("Fireside Chat" delivered by President Roosevelt on March 4, 1933.)

The New Deal

Immediately after his inauguration, President Roosevelt—FDR, as he was affectionately called by some—launched a massive revitalization program called the "New Deal," which he summarized as a "use of the authority of government as an organized form of self-help for all classes and groups and sections of our country."[34] With cotton prices having dropped to just 5.1 cents per pound—the lowest since 1897—such a program could not have come at a better time for southern cotton producers, tenant farmers, and sharecroppers.

Expressing his intent to move the country out of the Depression by creating jobs for the jobless, Roosevelt said he was not interested in "the dole" but instead was deeply committed to preserving the pride of American workers, whom he called "forgotten Americans—blacks, women, immigrants, and others suffering from poverty—who looked to the New Deal as a way to regain control of their lives."

State's Robinson Sold New Deal in Congress

Under this headline, Alex Daniels wrote that when Roosevelt took office almost three and a half years into the Great Depression, he inherited a catastrophic economic crisis that President Hoover and Congress had been unable to halt. But Roosevelt, America's thirty-second president, came into office with the New Deal. He also had a Capitol Hill champion on his side—the formidable Joe T. Robinson, sixty, of Lonoke. He was a former Arkansas governor (for just fifty-five days) and a former vice presidential candidate (with Al Smith in 1928). He had served in Congress for thirty years and, after Roosevelt's one-sided victory in 1932, was now the Senate majority leader.

With Robinson running interference for the new president, Roosevelt's first priority was to stabilize the banking industry—thousands of banks had closed in 1931 and 1932. On Monday, March 6, just two days after being sworn in, Roosevelt "declared a four-day bank holiday, called Congress into emergency session, and pressed lawmakers to pass a law that would allow branches to quickly reopen without attracting a run by nervous depositors. When asked by Vice President John Nance Garner and Speaker of the House Henry Rainey of Illinois how quickly Robinson could push the bill through the Senate, he replied that he could do it by that night.

"They broke out laughing," Daniels said, "but they had underestimated the man the press had nick-named 'Scrappy Joe.'" After it had passed quickly through the House and Senate and within six hours of opening debate, Robsinon sent the new banking bill to Roosevelt for his signature. As specified in the new measure, "Banks could sell preferred stock to the government to gain liquidity and were allowed to reopen after an inspection of their balance sheets by federal investigators. Banks began to reopen Monday, March

13. Rather than hoard their cash, depositors—their faith in the system reestablished—took their money to the banks."[35]

FDR's First One Hundred Days

Also a high priority in Roosevelt's New Deal strategy after his inauguration was to send Congress a series of laws, which were quickly passed. Taken together, they "dramatically increased the role of government and provided a safety net for people out of work or without homes."[36] During those one hundred days, Congress passed the Civilian Conservation Corps Act; the Agricultural Adjustment Act (AAA); the Economy Act, which reined in government spending; federal insurance of bank deposits (later to become the FDIC); and the Federal Emergency Relief Administration (FERA) to provide relief or for wages on public works.

• **The Civilian Conservation Corps (1933).** As envisioned by President Roosevelt, the two purposes of the CCC was to provide outdoor employment for jobless young men between the ages of eighteen and twenty-six to build state parks and bridges, clear trails, mark historical sites, clean out streams, and develop roads. In fact, the state park legacy of the CCC in Arkansas includes Crowley's Ridge near Paragould, Devil's Den near West Fork, Lake Catherine near Hot Springs, Mount Nebo near Dardanelle, and Petit Jean—Arkansas's first state park near Morrilton.

Given such nicknames as "Roosevelt's Tree Army," "Tree Troopers," and the "Three-Cs Boys," the CCC personnel lived and worked in 106 camps in Arkansas under military conditions and received cash allowances of $30 a month, with mandatory $25 allotment checks sent back to their families. Despite President Roosevelt's objections, the CCC was abolished by Congress in 1942.

• **The Agricultural Adjustment Act (1933).** The AAA was intended to reduce surpluses in cotton and livestock and to raise prices in an attempt to return them to the "parity" levels before World War I. In return for plowing under about 25 percent of their new cotton crops, for example, farmers would receive parity payments from the AAA ranging from $7 to $20 acre, based on the average yield per acre of the crops that had been taken out of production.

Although President Roosevelt thought that farmers were expecting legislation that would increase farm prices, Brooks Hays, the Democratic committeeman of Arkansas who was later elected to Congress, was dismayed to discover that the New Deal was of far greater benefit to landowners than it was to tenants and sharecroppers. Secretary of Agriculture Wallace had "failed to understand the roots of the agricultural crisis, particularly in the South," said the historian William J. Atto. "Sharecroppers, with little or no political influence, were not part of the conversation in the shaping of the Agricultural Adjustment Act of 1933."[37] As a result of farmland being taken out of production, thousands of tenants and sharecroppers were forced off the farms and were suddenly homeless.

The Agricultural Adjustment Act also resulted in placing Secretary Wallace into the difficult position of persuading farmers to plow up about ten million acres of cotton and to slaughter some six million squealing piglets[38] and 220,000 pregnant cows in 1933 alone, which they did. As a result, gross farm income increased by 50 percent from 1933 to 1935 but many farm families were devastated.

The act, which was a factor that provoked sharecroppers to form the Southern Tenant Farmers Union (STFU) in 1934, was declared unconstitutional in 1936 by the U.S. Supreme Court because "a majority of the judges said it was illegal to levy a tax on one group, the processors, in order to pay it to another, the farmers."[39]

The "Second" New Deal

When it became clear that more Americans needed federal relief assistance, the Roosevelt administration passed additional legislation between 1935 and 1938. Among those was the Works Progress Administration, established in 1935.

- **The Work Progress Administration.** Considered the most far-reaching of the public-works programs, the WPA is remembered for projects ranging from building schools and community centers and paving roads to offering programs for those hit hardest by the Depression on how to sew clothes and can foods. For their efforts, workers were paid fifteen to ninety dollars a month, depending on the job.

As noted in a report published by the Arkansas Historic Preservation Program in Little Rock, new WPA construction in Arkansas through 1939 included 773 buildings, 297 schools, 81 gymnasiums, and 34 stadiums. Of particular interest to Razorback football fans, WPA workers built the 13,500-seat predecessor to Reynolds Razorback Stadium in Fayetteville in 1938. Other construction noted in the report included 16 hospitals, 4 fire stations, 153 other public buildings, and 49,731 sanitary privies and thousands of miles of county roads. As the wartime economy boomed and with increasing employment in the defense industry, the WPA was closed in 1943.

- **Cultural contributions of the WPA.** As observed by Ben Johnson III, benefits of the New Deal programs were often overlooked, such as their cultural impact on the state by encouraging interest and respect for local history, art, and folk traditions.

The Federal Writers' Project (FWP), for example, was established in July 1935 as part of the Works Progress Administration. Unemployed and underemployed writers developed local histories, travelers' guides, and cultural narratives, particularly those relating to oppressed groups such as Native Americans and African Americans. Among the publications produced by state historians and writers was "Arkansas: A Guide to the State" (1941), which provided valuable information on different regions and the larger cities in the state. Other noteworthy FWP contributions included interviews with former slaves living in Arkansas that resulted in an oral history project called the "Slave Narratives."

Still another significant contribution to the cultural life of Arkansas was an adult education program "that sent instructors into white and black communities across the state to teach basic reading, writing, and mathematics. One group of teachers targeted the tenant farmers of eastern Arkansas, providing additional classes in home-related topics, including gardening and proper sanitation."[40] Other outreach made possible by WPA programs included the arts, such as painting and theater. Not to be ignored are the WPA's financial contributions to the state: $117 million in federal funds and $36 million from local and state supporters.

• **Social Security Act.** Regarded by President Roosevelt as the New Deal's "cornerstone," the Social Security Act was pased in 1935 as a safety net for senior citizens, as well as for those who were unemployed and disabled. The government program is funded by current wage earners and their employers.

New Deal not without Its Critics

Still, as pointed out by Alex Daniels in "State's Robinson Sold New Deal in Congress," the New Deal was not without its critics. During a 1936 Liberty League dinner, for example, former New York governor Al Smith compared the New Deal to a Soviet-style regime. "There can only be one capital, Washington or Moscow," he said. "There can only be the pure, fresh air of free America, or the foul breath of communistic Russia."

Jeannie Whayne was quoted in the same article as saying that while she doubted that the New Deal saved Arkansas from bankruptcy, "it certainly helped. As an agricultural economy rather than an industrial economy, in some ways, we weren't hurting as much as the rest of the country. But the agricultural program put a lot of money into the hands of a lot of farmers. In

Northwest Arkansas, the New Deal provided jobs where there had been none. At its best, it put a lot of money into the hands of a lot of ordinary folks."[41]

A more critical view of the New Deal is found in a recent study by Amity Shlaes, who argues that the FDR policies actually made things worse by imposing burdens on business enterprises that would have adjusted and restored a market economy free of federal social welfare programs. (For additional reading, see *The Forgotten Man: A New History of the Great Depresison* by Amity Shlaes.)

Perhaps the best question to ask concerning the effectiveness of the New Deal is "how would Arkansas have suffered had there *not* been a New Deal?" Donald Holley answered it this way: "Federal agencies saved homes and farms, provided work, built public facilities and improved the quality of life. The New Deal brought hope to people who had lost confidence in everything, including their country and themselves. Looking back from today's perspective, those who remember the Depression believe Franklin Roosevelt saved the country."[42]

Establishment of the Dyess Colony

William R. Dyess, a plantation owner and state administrator of the Federal Emergency Relief Agency and the Works Progress Administration, is credited for coming up with the idea of the Dyess Colony, which was seen as a way to convert rich delta land in Mississippi County into productive farmland that would provide residents with a good source of income. After hearing the proposal, Harry Hopkins, special advisor to President Roosevelt, came up with $3 million in federal aid to purchase 16,000 acres of what Dyess had called "swampy and snake-infested" Mississippi County bottomland, and convert it into a "resettlement colony" that would provide homesteading families about 20 acres of land for cultivation.

After the Colony was established in May 1934, one of 102 towns founded by the administration of Franklin D. Roosevelt to help families get a new start after hard times, the first of about five hundred families began to arrive in August, who started the daunting task of clearing the land, cutting down trees, blasting out stumps, and plowing each of the twenty-acre plots to plant cotton, corn, and soybeans, and to provide pasturage for their livestock. In return for cultivating the land, each of the families was given a home, a mule, a cow, groceries, and farm equipment. They were also expected to pay back their loans as soon as the crops sold.

Even using its own money—called "scrip" or "doodlum"—Dyess eventually grew into a community that contained three hundred barns and farmhouses, an administration building, a community bank, a beauty salon and barbershop, a blacksmith shop, a café, along with a post office, service station, print shop, theater, hospital, cotton gin, library, general store, schools, a sorghum mill, and a newspaper, the *Colony Herald*.[43]

One month after the death of Mr. Dyess in an airplane crash, the project was incorporated in February 1936 as the Dyess Colony. On June 9, 1936, First Lady Eleanor Roosevelt paid a visit, delivered a speech at the administration building, had supper at the Dyess Café, and shook hands with locals for several hours. At the time of her visit, there were about 2,500 residents in the community. In 2005, the population was about 480. A reunion for former Dyess residents and their descendants is held each summer.

Sharecroppers and the Rise of the Southern Tenant Farmers Union

In describing the "squalor, ignorance and disease" in which Arkansas sharecroppers lived in the 1930s, Donald Holley wrote:

> They lived in rickety, weather-beaten shacks with cotton planted right up to the back door. The monotony of their diet—salt meat, cornbread and molasses or sweet potatoes—explained the high incidence of pellagra among tenant children; malaria and hookworms were also common. Whole families worked in the fields from "can to can't" from the time they could see the dawn until they could no longer see at dark. Even by depression standards the income of tenant families was below subsistence levels, rarely averaging more than $200 or $300 a year.[44]

"Browbeaten by generations of intimidation backed as needed by noose and fire," David Kennedy said, the majority of the tenants lacked both the courage and the means to protest the conditions in which they lived.[45] The seeds for change were planted, however, in 1934 when two young white Tyronza-area socialists and businessmen—H. L. Mitchell, a dry-cleaner and a former "cropper," and Clay East, a service-station owner, after being encouraged by the successful if short-lived establishment of an unemployment league to advocate tenant rights—came up with the idea that the long-term solution to rural poverty and dependency in Arkansas would be the establishment of a socialist organization. A similar idea—that a union be started for farmers—had been suggested by Norman Thomas, head of the national Socialist Party, while visiting the state in February 1934.

So it was with this in mind that in July 1934, Mitchell and East led eighteen tenants—eleven white and seven black—into the Sunnyside school house in Tyronza with the intention of forming the Southern Tenant Farmers Union. Understandably, the action was viewed as a new and serious threat to white supremacy.

"Reprisals were swift and savage. Riding bosses descended on STFU meetings with whips and guns, hounded and beat organizers, and inveighed against the pernicious influence of 'outside agitators,' including Norman Thomas." After being knocked off the platform by sheriff deputies, Thomas was told, "We don't need no Gawd-damned Yankee bastards to tell us what to do with our niggers.'"[45] In the aftermath of the organizational meeting, said Doug Smith, associate editor of the *Arkansas Times,* "STFU organizers and sympathizers were arrested and beaten. One white organizer was arrested on charges that included calling a black man 'Mister.'"[46]

The woeful condition of Arkansas sharecroppers had already attracted the interest of the National Socialist Party, but Nat Ross, the district organizer for the Communist Party of the United States, saw the seeds of revolution that "could convert thousands of sharecroppers into communists who could infiltrate the STFU."

Although the STFU was at first dismissed as a joke and never operated as a real union, reports of frequent arrests and beatings gained national interest and sympathy. But Arkansas historians say the union began to fall apart in 1939 when its communist members attempted to oust the union's socialist leaders—J. R. Butler, the president, and H. L. Mitchell, the secretary. As noted by the Associated Press reporter Kevin Freking, however, if Mitchell and Butler "were at odds with the communists, the records at the Library of Congress don't reflect it."[47]

Perhaps the STFU should best be remembered for its pre–Civil War era example of using the effectiveness of racial integration to achieve its goals.

> Its use of rituals, songs, speech patterns, and evangelism became a model for the southern civil rights movements to draw upon. Though these techniques had been used in varying forms throughout the nineteenth century to mobilize public opinion, the STFU's religious fervor was new. This combination of evangelism, practicality, and purpose would provide an operational example for later civil rights and women's movements, setting the stage for the next half of twentieth-century American life.[48]

After several name changes and affiliations with other unions, the STFU went out of existence in the 1960s. Yet, with its strong religious undertone, its integration of African Americans and women into its membership, and its nonviolent means, the STFU left a lasting legacy and set the stage for the civil rights movement almost three decades later. The STFU Museum, which opened in 2006, is located at 117 Main Street in Tyronza, Arkansas.

Arkansas after the Great Depression

As World War II drew closer and the economy began to improve in the late 1930s, increasing reliance would be placed on federal programs. That would lead to what Jeannie Whayne called "a mass demographic shift" that was partially created by the New Deal. With increasing attention being given to mechanization made possible by funds generated through New Deal programs, combined with the effects of the AAA, Whayne said, more and more tenants and sharecroppers were moving out of the state. "The next two and a half decades," said Whayne, "would witness a major transformation in the state, a transformation that could not have been foreseen or forestalled."[49]

II. Meet the Survivors

Celia Almyra Graham Acrey • DOB: March 11, 1914 • Sidon★ • County: White
Dorothy Cox Coston Ashley • DOB: May 7, 1928 • Wesson • County: Union
Billie Bagby • DOB: January 25, 1920 • Bexar • County: White
Kathryn Bajorek • DOB: December 28, 1937 • Coy • County: Lonoke
Bessie Blacknall • DOB: April 2, 1919 • Mount Morriah • County: Clark
Billy Hayden Blankenship • DOB: April 14, 1926 • County: Izard★★
Fred Norman Blankenship • DOB: July 3, 1916 • Wideman • County: Izard
Herman Block • DOB: February 12, 1926 • Monticello • County: Drew
Nancy Duffy Blount • Interview on mother, Leona Stith Duffy • DOB: April 30, 1911
Rufus Thomas "Skinny" Bone • DOB: August 29, 1917 • County: Izard★★
Louise J. Brant • DOB: July 25, 1922 • Ozark • County: Franklin
Ethel Kent Brockwell • DOB: September 9, 1907 • County: Izard★
Faye Hargis Anderson Byrd • DOB: February 12, 1910 • County: Izard★★
R. L. "Bill" Carter • DOB: April 16, 1925 • Rich • County: Monroe
Mary Leona Cartillar • DOB: January 10, 1921 • Smackover • County: Union
Doyle L. Collins • DOB: September 9, 1928 • Near Ash Flat • County: Fulton
Geneva Louise Beckett Cotton Collins • DOB: March 18, 1932 • Barber • County: Logan
Leta Tisdale Curry • DOB: July 28, 1919 • On farm • County: Hot Spring
William Edward Delamar • DOB: May 4, 1918 • Near Dalark • County: Dallas
Tennessee "Tennie" Lawhon Dover • DOB: June 27, 1898 • County: Izard★★
Spencer Lee Duffy • Interview on mother, Leona Stith Duffy • DOB: April 30, 1911
S. Loretta Echols • DOB: May 20, 1935 • Humphrey • County: Arkansas
Artemeze Edwards
Jean Edwards • DOB: May 18, 1929 • North Little Rock • County: Pulaski

Geneva King Emerson • DOB: April 14, 1935 • Grange, near Cave City • County: Sharp
Robert Eubanks • DOB: March 24, 1929 • Paragould• County: Greene
Beulah Lee McLeod Evans • DOB: June 21, 1921 • Walnut Ridge • County: Lawrence
LaVerne Williams Feaster • DOB: October 14, 1926 • Big Dixie • County: Woodruff
Jeffie Naomi Moser Felton • DOB: November 22, 1897 • County: Izard★★
Lucille King Havner Finley • DOB: 1903 • County: Izard★★
Robert Flannigan Sr. • DOB: December 25, 1923 • Monette • County: Craighead
Mary Eathel Freeman • DOB: December 13, 1920 • Okay • County: Pope
Herndon Frizzell • DOB: October 24, 1909 • County: Izard★★
Doyne Harold Gaston • DOB: November 24, 1922 • County: Izard★★
Vernon L. Glenn • DOB: June 29, 1926 • Lynn • County: Lawrence
Audie Tate Grigsby • DOB: March 13, 1908 • County: Izard★★
Ralph Emerson Hall • DOB: November 8, 1929 • Holly Grove County: Monroe
Zelma Louise Hamilton • DOB: October 19, 1925 • Moro • County: Lawrence
John Henry Owen William Harvell • DOB: September 12, 1909 • County: Izard★★
Menard Eulis Harvell • DOB: June 5, 1918.• County: Izard★★
Violet Hensley • DOB: October 21, 1916 • Alamo community • County: Montgomery
Georgia Hearn Hill • DOB: November 15, 1931• Near Gum Springs • County: Clark
Lois Taylor Horton • DOB: July 4, 1928 • Snowball • County: Searcy
Mary Frances Lovell Izard • DOB: April 16, 1938 • Near Bryant • County: Saline
Roscoe E. Jefferson • DOB: December 2, 1911 • Near Yellville • County: Marion
Grace Ferguson Johnson • DOB: June 4, 1915 • County: Izard★★
Rayburn Thomas Johnson • DOB: May 21, 1922 • County: Izard★★
Don Orville Knoll • DOB: October 16, 1933 • Near DeWitt • County: Arkansas
Reba Knox • DOB: May 9, 1935 • Family farm was near Waldron • County: Scott

Kenneth Guy Lacy • DOB: December 17, 1923 • Near Heber Springs • County: Cleburne
Evelyn Langley • DOB: December 14, 1919 • Near Guy • County: Faulkner
Willis Magby • DOB: August 1, 1920 • Beaton • County: Hot Spring
Suzanne Gross Marks • DOB: September 27, 1923 • Lonoke • County: Lonoke
Dulas L. Massey • DOB: December 25, 1913 • Scotland • County: Van Buren
Mamie H. Mays • DOB: October 13, 1916 • Near Malvern • County: Hot Spring
Marion Anderson McAnally • DOB: January 1, 1909 • Fendley • County: Clark
Rupert E. "Buster" Melton • DOB: January 1, 1909 • Cord • County: Independence
George Mertens • DOB: March 12, 1927 • Farm near DeValls Bluff • County: Prairie
Evelyn M. Coonfield Metcalf • DOB: April 11, 1935 • Eureka, Kansas • County: Greenwood
William Wesley Miller • DOB: December 6, 1920 • County: Izard★★
Howard Peyton Moore • DOB: December 27, 1914 • County: Izard★★
Willie Morris • DOB: April 3, 1924 • Jerome (near Dermott) • County: Drew
Alma Lorene Reed Neal • DOB: July 3, 1919 (deceased) • County: Izard★
Michael Francis O'Cain • DOB: 1892★★
Lloyd A. Perry • DOB: May 4, 1920 • On a farm near Guion • County: Izard
Violet Piercy • DOB: January 27, 1932 • Osage area • County: Arkansas
William Piercy • DOB: September 26, 1929 • On farm • County: Benton
Alma Pounds • DOB: August 8, 1914 • Farm near Jonesboro • County: Craighead
Opal Copeland Radford • DOB: April 8, 1909 • Near Jonesboro • County: Craighead
Lydia Moser Rider • DOB: July 5, 1916 • County: Izard★★
JoAnne Rife • DOB: June 28, 1926 • Bentonville • County: Benton
Paul Roberson • DOB: September 8, 1924 • Farm near Okolona • County: Clark
Phillip Wayne Rowan • DOB: January 7, 1896 • Near Amity • County: Clark★★★

Union H. Stoudamire • DOB: January 1, 1906 • Rural Arkansas• County: Jefferson

Marie Taylor • DOB: April 4, 1927 • Grew up near Manilla • County: Mississippi

James A. Thompson • DOB: May 19, 1928 • Cave Creek • County: Newton

Eva Smoke Wells • DOB: January 1, 1925 • Near Malvern • County: Hot Spring

Arwilda Whiteside • DOB: August 22, 1925 • Near Clarendon • County: Monroe

Charles Whitfield • DOB: November 25, 1906 • Near Bismarck • County: Hot Spring

Herron Thompson Whitfield • DOB: September 24, 1914 (deceased) • County: Izard

Albert M. Williams • DOB: March 3, 1912 • Farm near Wilmar • County: Drew

Lucille Rider Wilson • DOB: October 4, 1924 • Rural area near Ozark • County: Franklin

Floye Wingfield • DOB: June 16, 1916 • DeGray community • County: Clark

Jack Wood • DOB: October 11, 1918 • Farm south of Arkadelphia • County: Clark****

*All cities and towns are in Arkansas unless otherwise stated.
**Courtesy of the "Down Memory Lane" series by Betty Guthrie McCollum, Melbourne, Arkansas, and Sue Shell Chrisco, Sage, Arkansas.
***As told by his son-in-law John E. McCown Sr., LTC, U.S. Army, Retired.
****Combined with an earlier interview by Maggie King, daughter of Mr. and Mrs. Taylor King of Arkadelphia.

Photo Gallery

Celia Almyra Graham Acrey

Dorothy Cox Coston Ashley

R. L. "Bill" Carter

Mary Leona Cartillar

Doyle Collins (right) with grandson Rance Collins (left) and son Ace Collins after graduation ceremonies at Ouachita Baptist University.

Geneva L. Collins

Leta Curry

Beulah Evans

Bob Eubanks

Jean and Artemeze Edwards

Geneva King Emerson

Eunice and Willis Magby

LaVerne Williams Feaster

Vernon L. Glenn

Ralph Hall (second from right), with Mike Cates (son-in-law), Louise Hall, and daughter Janet

Zelma Hamilton (left) with her sister Zadie in 1930

Georgia Hearn Hill

Scarlet King (left) taking fiddle lessons from Violet Hensley

Mary Frances Lovell Izard

Dulas Massey

Evelyn Langley with granddaughter Lynita Langley-Ware, who is holding daughter Sophia

Mr. and Mrs. Cleovis Whiteside

Mamie Mayes

Mr. and Mrs. Rupert Melton

George Mertens

Mr. and Mrs. William Delamar

Mrs. Union H. Stoudamire

JoAnne Sears Rife

Mr. and Mrs. Fred Blankenship

Alma Pounds

Nancy Duffy Blount

Spencer Duffy

Four generations: Roscoe Jefferson, grandfather; Shirley Cox, daughter; Shannon King, granddaughter; and Scarlet King, Mr. Jefferson's great-granddaughter

Eva Smoke Wells

Charles Whitfield

Albert M. Williams

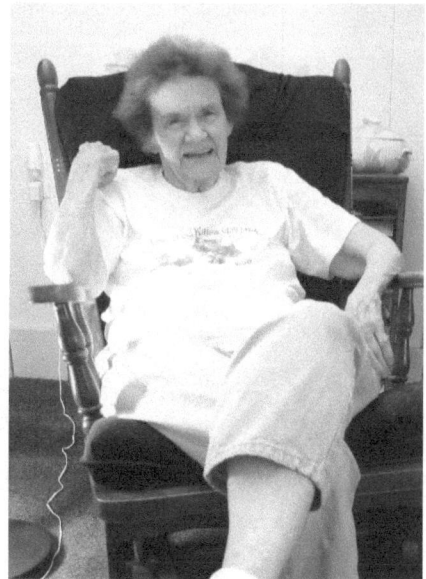

Floye Wingfield

Lucille and Virgil Wilson

Mr. and Mrs. Jack Wood

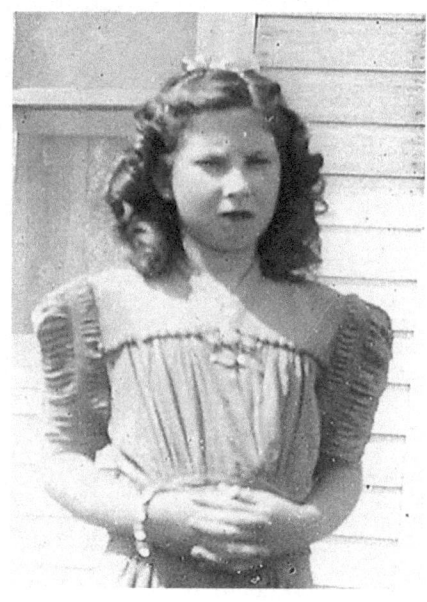

Evelyn Metcalf, at age 11

Paul Roberson

Marie Taylor

Lloyd A. Perry

Robert Flannigan, Loretta Echols, and her great-granddaughter Abigail Mitchell at an ASU basketball game

III. Stories of Survival: Arkansas Farmers during the Great Depression

How "Stories of Survival" Interviews Were Conducted

Using the "Stories of Survival" interview format (see Appendix), the interviews were conducted either face-to-face or by telephone. Although each person interviewed was asked the same questions, I was careful to go beyond their first responses by asking for more information: How did this affect you, how did you react? etc. Early on, I also learned not to interrupt the interviews when something was said that prompted a response from me—which usually stopped the subject's train of thought. In addition, I was not afraid of what some call "pregnant silences," those times when nothing was said, which can prompt more responses. Almost invariably, something of value would emerge from the person being interviewed such as "I just remembered something else that I would like to share." In other cases, they would send additional information.

Very important, I always, *always* asked at the end of the interview—as noted in Question 26—"Is there anything I didn't ask about? Memories? Favorite stories? Little incidents that didn't seem to be important at the time but have remained in your memory?"

All interviews were taped and later transcribed by the author. To ensure accuracy, a copy of each transcription was returned to the person interviewed along with postage-paid return envelopes. I cannot overemphasize the importance and effectiveness of this approach. Once the subjects realized that they could trust me, they usually were willing to share some very personal stories. Once the transcriptions were returned, I would then enter new information, deleting or correcting what was inaccurate. Finally, as I continued to conduct the interviews, I saw what worked and what didn't and would revise accordingly.

Although I am well aware that there are surely hundreds of "survivors" in Arkansas whom I did not reach, there came a time when I was increasingly getting basically the same responses. For this reason, I was unable to use quotes from each person interviewed. I sincerely hope they will understand. The name of each person interviewed or whose memories are included in this book, however, such as those in Izard County's splendid "Down Memory Lane," whose names are listed in the "Survivors."

Chapter 1

Earliest Memories of the Great Depression

BESSIE BLACKNALL

We started out with one room. Then a kitchen and a bedroom with a fireplace. Mother cooked on a woodstove. There were six persons living in the house—my parents, two sisters, and two brothers. I wished we had had a better house.

FRED NORMAN BLANKENSHIP

I wasn't old enough to worry about anything. I just heard my folks talking about losing money, how the banks were going down and all of that. If I remember right, and I think I do, my daddy sold a team of mules but hadn't taken a check to the bank so he didn't lose anything when the bank closed. He didn't have much money in the bank at that time but he would have had that much more if he'd gone to the bank after he sold the mules.

After you came in from the fields, cleaned up and had supper, how would you spend your evenings? Every family had several children. What I remember particular about it, my uncle lived about a quarter of a mile from us and a friend lived another direction from us. Almost every night families would come and sit down on our front porch. My Daddy would chew tobacco and the other fellows would dip snuff. He went to bed every night at eight o'clock. It had to be kind of a special person if he sat up past eight. He would get us up at five o'clock the next morning, too!

We played ball and a lot of games kids used to play, you know, like hide-and-seek, that kind of thing. People don't visit that way now. I would usually go to a friend's house on Saturday night and he would come to our house the next Saturday night.

Ghost stories and Aesop's Fables. My grandmother took the weekly *Kansas City Star* that came every Wednesday. There were two things that were always in the *Star*—a ghost story and Aesop's Fables. That night I would go to her house. Although I could read, I liked for grandmother to read the ghost stories to me because she could make them sound scary. I would have been disappointed if neither of those made the paper.

R. L. "BILL" CARTER

We done without. I remember one time we got down to where we had nothin' for bread. But we had some nubbins of corn out in our barn. We took a bucket lid and nail, drove holes in it and went out there and got some of those nubbins and rubbed it across there to make meal out of so Mama could make corn bread. Now that's poor! [chuckling softly] There was one family that lived in our community that in the spring of the year when the leaves started budding out, they cooked them young leaves like they were turnip greens. I don't remember them saying what kind of leaves they was.

Rocks in the beans. Back in 1930 I remember the Red Cross giving us sacks of beans that had rocks in them. It was just like they had gone out there and piled up along a rocky place and loaded the rocks, which were the same color as those beans. You'd have to be real careful if Mama had failed to catch the rocks when she was cooking them.

How Dad got his smoking tobacco. Dad used to smoke in his early years. I remember he didn't have money to buy tobacco. He would send us kids out on the railroad track when the passenger train had gone by an we would pick up cigarette butts. He would take them, get the tobacco that was left in them, and smoke. He had a dream one night that the Lord had come back and he wasn't ready to go. So he went down to my grandpa's house, which wasn't too far behind our house, and got saved. He left his can of tobacco on the shelf there and never did go back to get it. He stopped smokin' that quick.

MARY LEONA CARTILLAR

Fighting mosquitoes. Sometimes they were really, really bad. Daddy would build "smokes" in the house to get rid of the mosquitoes. He would start the fire and then he'd smother it down to a smoke and it would be in a bucket. He'd set it in the house to get rid of the mosquitoes, because they were horrible, especially around Hickory Ridge—that's where they raised rice. And it was awful.

WILLIAM EDWARD DELAMAR

Black or white? My daddy sharecropped some and we rented some. There are two sets of Delamars where I was raised, one was black—my family—and one was white. And the black and white come up together. Anytime someone said anything about the Delamars, I would ask, "Black or white?"

[Author's note: Mr. Delamar is African American.]

ROBERT EUBANKS

A typical supper. It usually consisted of corn bread and milk, and sometimes we had full meals with various cuts of pork, green beans. We seldom bought bread, which we called "light" bread. We even had biscuits in our lunch at school. A typical lunch would be Mama cooking enough sausage for us to take. We would split a biscuit and put a cake of sausage in it, and that would be our lunch. Sometime a kid would have a sandwich made out of light bread and we would swap because we liked that light bread.

BEULAH LEE MCLEOD EVANS

Memories of her mother. Viola McLeod was a small person, but she had more guts than anybody nearly. She was a very religious person. I remember seeing her a lot of times in the back yard walking around under those big oak trees and her mouth moving. She was praying. She prayed us all through. She done all the cooking. And part of the time she'd even help out milking the cows. We milked the cows by hand. A lot of times she'd help out with that. And then she'd always clean up the separator after we was through separating the milk. And do all that work. And then when she was finished, she'd go to the field and work and then fix all the meals. I never remember dirty dishes stacked up at our house. I don't remember us helping with the dishes. I remember Mother doing them.

VERNON L. GLENN

A typical day on the farm. We would get up early and while my mother was cooking breakfast, we fed the mules and fed and milked the cows. Then we would go in and have breakfast. After that, we would get organized to go out on the farm to start plowing, chopping cotton or whatever else was to be done that day. If we were close enough to the house at noon, we would go back to the house for lunch. Otherwise, my mother would put lunch for us in a bucket and one of us would go back to the house to get it and bring it to the field. The mules would eat grass or hay, which we had brought with us that morning and would drink from the nearest water hole or ditch.

VIOLET HENSLEY

I knew it was hard times. When I was ten or twelve years old, we didn't have no money to buy anything. My dad drank coffee and he bought coffee. He chewed tobacco—he raised tobacco. We didn't touch his tobacco patch. We raised corn, cotton, peanuts (for ourselves and our animals), Irish and sweet potatoes. We had vegetables in the summertime, whatever we needed—peas, beets (but not carrots, they wouldn't grow in that slatey dirt). They made them little ol' scrawny things. We had to catch rabbits for something to eat. We also trapped quail.

After the chores were done, we played. Sometimes we'd get a board and knock down bumblebees. We didn't have a lawn. We swept it with a brush-broom. When it got dark, I would play the fiddle. My dad was always making or playing a fiddle. The old porch was 1 x 6's and we would saw the fiddle down through the cracks in the floor.

My little mule "Cricket." When I was sixteen, we bought a little mule and called her "Cricket." She could jump the moon and swish the stars with her tail. She could pick the soda out of the biscuits and never mess the crust. She wouldn't kick me. I could crawl under that mule's legs and I guess I just had a way with animals. I could harness her up and plow it all day long. I would whisper to her. She really minded me. We had a piece of land across the creek. One time a neighbor caught me talking to my mule and asked who I was talking to. I told her I was talking to my mule.

How to make your own toothbrush. A toothbrush was a little piece of little limb off a black-gum bush. You would chew the end of it until it was soft then you would brush your teeth with it. That was our toothpaste.

MARY FRANCES LOVELL IZARD

We lived in hill country. Much of the land around the home place was "new ground" or recently cleared and still had a few large stumps that had not been removed. (My older sister referred to it as "stump and sprout land.") Also the soil is thin in the hills and more susceptible to drought. The area was also blessed with an abundance of quartz rocks which made it almost impossible to keep our tools sharp.

Description of our one-room house. The "little house," as it was referred to, had one square room which contained two beds and a couple of straight chairs (as opposed to rocking chairs). There was room for little else. There were no closets, maybe a small "chest of drawers" to hold folded clothes. Other clothes were hung on nails that were driven into the walls. The house was built with uncured rough boards bought from the sawmill or our own logs cut from our land, which the sawmill cut into planks. The house was not underpinned but was built on pillars of large local rocks.

There was another small room attached to the back of the square room. It had a sloping roof and was referred to as a "lean-to" room used for the kitchen and eating area. It contained a wood-burning cooking stove, a cupboard, and a table with chairs for the adults and a bench behind the table for the children.

There was always a small shelf on the wall that contained the water bucket and dipper (which everyone drank from), a small metal pan in which to wash your hands, lye soap that my mother made, and a flour-sack towel hanging on a nail on the wall. The table and the bench were built by my father out of rough planks. The table was always covered by an oil-cloth. The older children sat on the bench behind the table. The smaller ones might require a Sears & Roebuck catalog in order to see the table top. There was no special furniture for the baby. If the baby was not sleeping, the mother held it while she ate. Outbuildings included a barn, a corncrib, a pigpen, a chicken house, a smokehouse, and most important of all, an outdoor toilet.

KENNETH GUY LACY

Keeping out the cold. Mom had tried to paper the walls with newspaper, cardboard, and pages from catalogs, using flour paste, trying to keep out as much cold as possible. In the summertime, we could hear mice and see their forms between the walls and paper, eating the paste. We would put our hands on top of the paper where they were underneath and mash them with our hands.

In the winter, we piled as many quilts on the bed as we could get, hardly able to turn and would still be cold. During snows, we would wake to find snow on top of the quilts that had filtered through the cracks. The floors were bare with cracks between each board through which we could see the ground beneath. Mom kept them clean by scrubbing them with sand and hot, soapy water. Sometimes a little lye was mixed in.

We had no electricity, only a kerosene lamp and light from the fireplace in the winter. The only running water we had was what we would run outside and draw up from the hand-dug well. It seems we had a lot of company that came to spend the night. Most of the families were large, having several children. I remember Mom would make down pallets on the floor for the kids. It was just the way of life and was accepted as such, our not ever having known anything else.

WILLIS MAGBY

Hard times. I remember all of it. I remember once a man saying, "Well, you lived in the country, so you had plenty to eat." I asked him, "Then why was I so hungry all the time?" There were times when we would have very little. If we had beans and corn bread, we's a'doing pretty good. It wasn't only my family that was that way. That was the rule rather than the exception.

RUPERT E. "BUSTER" MELTON

A typical evening after coming in from the fields. Well, us younger folks would play checkers to dark and after sometime. Another thing, I had some pretty good dogs back then and I possom and 'coon hunted. I'd skin 'em and sell their hides, sometimes fifteen or twenty dollars a hide. That was during World War II. After we'd skin them, sometimes we'd eat 'em. If we didn't want them, we'd throw 'em away.

MAMIE H. MAYS

Some people had a picket fence but we had a plank fence—four planks high with about six inches between each plank. Me and my brother would climb up on that fence on Saturday afternoon because the Masonics—the Masons—met at Lono. Our daddy was a Mason. The old people would drive their horses up to the house and talk to me and my brother. We'd sit on the fence and talk to them.

What do you remember about your mom and dad? Mother was older than him, but she was a virgin when they married. That's the reason he married her. She taught him how to add and subtract but they didn't make him go to school.

What was a typical day on the farm? Get up early! When they would holler "Breakfast!" The kids got dressed, washed their hands and faces, combed their hair before they got to the table, then had their breakfast. We'd usually have salt meat or ham or sausage Mama had canned, fresh eggs, sorghum molasses. When somebody killed a hog, they would peddle the meat out into the community.

After breakfast, we would go to the fields until she called us to come in for the noon meal, usually whatever was left over from breakfast. We'd rest a while and then go back to the fields.

Then we would come back, take our baths, all in the same water. The girls would bathe first, then the boys would bathe because they were dirtier than us. After that, if it was wintertime, we would sit around a fire and tell tales. Mama would always tell us a Bible story, saying the Lord would take care of us through the night. She was the main one. Then we would go to bed.

GEORGE MERTENS

We were poor but didn't know it. By the time I started to school, my dad had cleared most of the forty acres and had about fifteen acres of fruit trees and two acres of strawberries. We had hogs, chickens, milk cows, beef cows, work horses, along with an old Fordson tractor. We always had a big garden and numerous truck patches of sweet corn, potatoes, sorghum, etc. My mother was a great cook and hard worker who usually put up four hundred to five hundred jars of produce—peaches, pickled peaches, blackberries, plums, green beans, kraut, tomatoes, butter beans, and she also canned beef—for the winter.

A typical day on the farm. That all depended on the season, of course. There were always daily chores such as milking cows twice a day, feeding chickens, gathering eggs, feeding hogs, horses, and cows. In the spring and summer it was working in crops, plowing, hoeing, and in fall gathering crops. There was always something to do even when it rained and was too wet to work in fields. You may be picking blackberries or picking peanuts off vines stored in the hayloft. If we were lucky on those days, we got to go fishing in the White River, which was nearby, or hunt mushrooms after a spring rain.

Family life: The thing I remember about family life we were a close-knit community with a lot of uncles, aunts, and cousins and we would visit kinfolks and spend the night, kids on pallets on the floor, and listen to the men folks tell their favorite stories.

Bad times for some. Some of our neighbors, some of them kinfolks, didn't fare as well as we did. Some didn't own land and were day workers in the timber industry or farmhands—low-paying jobs, to say the least—and those that did own farms for the most part were not as succesful as my dad. He was always busy, driving school buses, farming, hauling logs, butchering hogs, and supplying a local meat market in exchange for groceries.

WILLIE MORRIS

Dad didn't want to die with his shoes on. I was just two years old when my dad was killed. He was a timber cutter and didn't get away quickly enough when a tree fell on him. He wasn't killed instantly but he had enough life in him to tell tree cutters to pull his shoes off before they cut the tree away from him. He didn't want to die with his shoes on. I don't know why.

LLOYD A. PERRY

We lost the eighty-four acres because we owed $350 for it. We couldn't find any way to earn the money to even pay the 10 percent interest on the loan. We had a neighbor who also lost a farm—four hundred, maybe four hundred fifty acres and the buildings, including his home, because he couldn't come up with the 10 percent interest on his loan, either.

My single mother and three sons, one of them almost a baby, had to find another place to live. My uncle had a place he said we could use if we boys would pay the rent by cutting the brush from eight of the forty acres of land. Soon, however, he was about to lose all his farming equipment because he couldn't pay the interest on the $117 loan he had on it. So he told my brother and me that if we could come up with the money for the 1 percent interest, we could try to make a cotton crop on his land. I told him I'd get a loan and do it. I was seventeen years old then, and my brother Lois was eighteen months younger than I was. He laughed and asked me if I thought anyone would let a seventeen-year-old sign for a loan. He really didn't think we could do it, but it was going to be lost anyway.

I went to the local banker and told him what I wanted. He actually let me sign for the loan I needed. The year was 1937. The government was paying farmers not to plant all their cotton acreage, and we used that payment for seed and fertilizer. It was a big job for two young boys, and lots of decisions to make. When we disagreed on something, the one who could whip the other was the boss. That fall, cotton went up to 10.5 cents a pound. We'd made six bales on ten and a half acres, a really good crop for this country.

We saved up all our bales as ginned and took them to Batesville for selling. Then we came back by the bank, paid off the entire loan, and still had lots of money left. The banker's mouth fell open. He was so surprised, he made an exclamation using God's name. Two boys who never had a dime to spare before were rich by local standards. The neighbor men were astonished at what we'd accomplished.

JAMES A. THOMPSON

We had lots of bad weather over in the Cave Creek area. But we didn't pay much attention to it. I was just a little boy—five or six years old. My grandpa just kept on plowin' corn. He said he would quit someday. Some people would say, "Oh, this is a bad year. Look at the drought we're having!" Well, it wasn't any worse than it was two or three years prior to that. But he just kept on plowin' corn. It didn't bother him any.

I'm sure it was very hard in Newton County because nobody had any money hardly. Maybe your neighbor was a little bit better off than you were. Just live off of him. A lot of people went to Bentonville during that time. It's a fruit country and most most of them went up there to pick apples. They would stay up there for maybe six months and then they would come back home.

EVA SMOKE WELLS

Hearing my folks talk about President Herbert Hoover. How bad he was, and how they hoped that someone else would get in [the White House] and do something good for the country. There was a joke about "Hoover britches"—they didn't have any pockets in 'em because you didn't have any money.

But mainly I remember that daddy would take eggs to the local store. We didn't have a car, so he would walk. He swapped eggs for coffee, sugar, tobacco, and flour. When the WPA came along, all the neighbor men got a job but my dad didn't. Just another example of why we don't think much of govenment projects or promises.

In a way the Depression was tough. We had to worry about getting the necessary things we had to buy. As far as food was concerned, however, I don't ever remember our being deprived. Except sometimes they might run out of flour and wouldn't have any flour for breakfast for biscuits, but we could have had corn bread, which was good. That was because they always raised enough corn to take it to the mill and have it ground for cornmeal. There wasn't a flour mill around here and they didn't raise wheat.

Cat-head biscuits. My mama used to make biscuits, which my husband got to calling "cat heads." She kept a pitcher on the cabinet and when she got through making biscuits, she would take one little chunk and put it in the pitcher. That way, it just sat there until it got fermented. Next time she used it, she would put buttermilk in with it. The biscuits would swell up like they had yeast in them. And right in the center, she would have the grease in her pan and she would put one side down and then turn it over. And right in the center where her finger was was something that looked like a little dimple. And that was her trademark. She had the brownest crust you ever ate. Talk about good!

ARWILDA WHITESIDE

I remember wanting a piece of liver that cost fifteen cents a pound. "We had a fifteen dollar quota we could use each month. We had gone over the limit and I couldn't get that fifteen cents to buy that liver. But when we were married, we owned our own little house. We had chickens, ducks, hogs, everything. My husband got out and planted one of the most beautiful gardens in the community. We were young, and our furniture was made out of orange crates and apple crates—and they were beautiful. My family and my husband's family and all the children were just like sisters and brothers. And we were happy.

CHARLES WHITFIELD

We growed up right close together. Seemed like I was about twenty-three when we married. She must have been about nineteen or twenty. I've heard her laugh about us waiting to get grown. We went up to Billy Shuffield's house and got married. George Cook and Billy Thomas carried us right up here. We didn't have no car, so we went afoot. They carried us up there and we came on back home. We didn't have any trouble a'tall. She was helping me when she was able. She was sick a lot. She'd help me hoe cotton and pick cotton some. In her later days, she didn't try that much. She always had somethin' cooked for me.

When typhoid fever was bad. I had six brothers and three sisters. I had one sister who was always contrary. They couldn't get her to take no medicine. She'd take her medicine but when you turned your back, she'd put it under her mattress. There were two of the little boys that died from typhoid fever when she died.

Logging accident paralyzes brother. I had two brothers younger than I was, Henry and Herman. Henry and dad was cutting trees. They cut one of them long pine trees that started to fall but set back. Some said Henry called his dog and started to walk off under the tree. When the tree started to fall, Dad hollered at him but Henry run the same way the tree was falling and it hit him across the shoulders and back. It broke his back into pieces and paralyzed him. He lived a year or two but the bed sores killed him. Herman, my other brother, was out in Arizona working in the mines and got killed there. I give him the money to go out there and I paid to bring him back.

We always had plenty to eat. We had butter, molasses, bacon; sometimes we would have ham. I remember one year, I killed a big hog. It was nice weather. This was a big hog and it was fat. I didn't have no smokehouse. I had one room in the house I hadn't ever put a floor in . And I put the meat in a big box, buried it in salt, and set it in there. It turned off warm and hazy and my hams and my shoulders soured. I had to throw it all away. Big ol' hams that was all I could carry.

LUCILLE RIDER WILSON

We were a strong family unit around my father, who made a team of us. The purpose of the team was to work, to produce so that we could survive. He was a genius at pulling the family together. We were Riders. And as Riders, we were very special. Whatever the family goals were, they were mine. My daddy didn't hunt, he didn't fish. He worked. Mama worked because she had five children. No running water, no automatic anything. She had an open kettle where she boiled the clothes. We didn't have a good well, so we had to carry water a half mile from the creek. Because we had to work, we didn't play anything. Mostly, I don't know how to play even now.

FLOYE WINGFIELD

Cracks in the walls. When we moved from the DeGray community to the Joan community where we are now in 1926, there were some cracks in the walls of the house that was over an inch wide. You could cool off in the summer a little bit but there wasn't much you could do in the wintertime. That really got rough so you just put a toboggan on your head and got under all the quilts you could find.

We didn't put mud in the cracks because Mama wanted wallpaper. She and all the neighbors were saving their empty one-hundred-pound tow sacks. She washed them, straightened them out, and saved them in a stack. And when she got enough to cover the walls in this house, she put it on instead of canvas because it was thicker. She nailed all those tow sacks up, but it was another year before we had enough money to buy wallpaper.

Fire destroys house in 1929. I was at the school in Joan when it caught fire. The sheriff had reported it but by the time the fire department got out here it was too late. I will always remember the sound I heard when our house fell in and I knew that was my home that was gone and I had nothin' Everybody came to console us. It was a sad, sad day. But the sun was pretty and bright and it was nice and warm, so we were just fine.

By the time night came and Mama and Daddy had gathered their wits a bit, they decided which neighbor each child would stay with that night. Things got to be lookin' a little better. We'd spend the night with our friends and go to school the next day.

The vacant house across the road. Mama and Daddy told us what we were going to do. There was a vacant house across the road from here. We could live there until we could rebuild the house where we were. Then Daddy made another big announcement. He didn't have but two thousand dollars in insurance. We knew we couldn't build much of a house for two thousand dollars and have enough money to get us some shoes and have enough to live until he could make a crop.

Neighbors came to the rescue. They said they would volunteer and do the work. They started right away. The neighbors cut down the trees, put up siding, got donations from merchants in Arkadelphia to help out, brought home dishes, clocks, all those things we needed. We moved back into the house on January 1, 1930. It may not have been very much, but it was home and we were very, very happy.

Chapter 2

Sharecropping in Arkansas

HERMAN BLOCK
We sharecropped for different white farmers. We lived on a white farmer's plantation, where we worked his land. Basically, sharecropping worked the same way. A few farm owners settled up at the end of harvest season. If you worked their land, they furnished you all of the seeds, etc., and you purchased your goods, food, etc., from the general store on account. At the end of the harvest, the farmers took what you owed at the store from the money you made on the farm. That left very little, and the cycle would start all over again. If we made ten bales, the farmers would get five of them and the sharecroppers would get the other five.

LaVERNE WILLIAMS FEASTER
We lived on the Dixie Plantation near Cotton Plant. The plantation was a town in itself that I would guess covered three hundred to five hundred acres. My dad was the director of the plantation "commissary" store—where the tenants were required to make all their purchases—clothing, food, medicine, and home supplies—and he also rented the farmland. The rental land started at forty acres but increased according to the size of the families. At any one time, there would be an average of thirty-five to forty families living on the plantation. They raised crops and a percentage of their crops paid for their production. Mr. Clarence Gordon, who was the owner of the Dixie Plantation, had a maid and a butler. He was a nice man.

As director of the local cotton gin and store, Daddy also made loans to farmers. This included seed, farm equipment and food. Daddy kept records. When farmers brought their cotton to the gin, Daddy processed it into bales and weighed the seed and cut cotton samples and measured the length. The length determined the amount per pound that each bale of cotton was worth. He would determine the farmers' pay for the cotton, subtract what they owed the store and then would pay them.

Seldom if ever did families make much money from their crops. They paid their bills at the store and then began another list of needs charged to them at the time of crop harvest. Many of the families were on welfare and received commodities and other government assistance. All of these people worked, raised crops, yet they only made enough to live on during harvest time.

The plantation owners took care of the families' needs—health care, food, etc. Yet many of these families had deductions taken from their crops to pay for everything the owners paid for. The government commodities were also managed by plantation owners.

Many times farmers did not get enough money to pay their bills and have money left for family needs. When families did not work like the manager thought they should, he would punish them by not giving them the commodities they needed even though many of the sharecroppers may have been sick or needed the commodities to care for their families.

The plantation owners used every opportunity they could to keep the families healthy enough to work yet poor enough to be dependent on the system. The midwife did not have to work in the field when she had a baby due, yet families had to depend on the plantation boss for medical needs. One example: Sometimes women would wear pillows and act pregnant to get special care and opportunities. The owners wanted the families to continue to provide workers.

ALMA POUNDS

My first encounter with a black person. The sharecroppers on our farm were all white. I hadn't seen a dark person, a Negro, until I was about four. The first time I saw one, my sister and I were sitting on the front porch and we went screaming around the side of the house. Mother had to keep us until my granddaddy came by. This Negro was sitting in the wagon and he came up to the front outside the fence. Granddaddy came over behind him to find out what was going on. He made us go out and shake hands with the Negro.

MARIE TAYLOR

As sharecroppers, we may have been better off during the Depression than many others were. First, we had a house furnished rent-free. Second, we always had a pump on the back porch or near the house so our water was free. Third, my dad and two older brothers could cut wood for heating and cooking off the landlord's property—a least we could in the early years. When the wood became scarce, Dad bought a coal-burning stove in about 1936. Then, of course, we had to buy the coal for heating the house but still used wood for cooking. So we didn't have utility bills other than coal.

Fourth, our landlord had a commissary and we could buy groceries on the credit when necessary—just basics, such as sugar, flour, coffee, etc. Salt pork was the only meat he carried, I believe. Prices in town were cheaper, though. So in the fall when we had a little cash, Dad would stock up in town on some things; he might buy two hundred pounds of flour, one hundred pounds of sugar, a stand (fifty pounds) or two of lard and probably several pounds of beans. Then when those ran out, we could buy from the commissary on credit until the next fall. Dad took corn to the gristmill to be ground for cornmeal.

Sharecroppers cheating on landlords. I've heard stories that some landlords cheated their tenants, especially the blacks who didn't have enough education to keep up with their accounts. It wouldn't surprise me that some did cheat. But I heard my dad say more than once that our landlord, Mr. W., was honest and if he told you he would do something, he would do it.

There were some people who would cheat the landlord if they got a chance. Our landlord had several families, black and white, who lived on his place year 'round, then in the fall, a lot of other people came to the East Bottoms to pick cotton.

Dad told us of one man who cheated our landlord. Dad was the weigh boss, weighed everyone's cotton and kept records of it. Saturday afternoon was payday or "settling-up time." One Saturday, when they were ready to pay a certain man, he disagreed with Dad's figures on how much cotton he had picked—I think about one hundred pounds. [That was] more than Dad's figures, which would have been seventy-five. Mr. W. asked Dad if he might have made a mistake. Dad said he could have but didn't believe that he had because he went over all figures very carefully. Mr. W. said give him the benefit of the doubt and pay him by his figures.

Well, the next week, the same thing happened again with the same man. Mr. W. said, "Frank, pay him by his figures, then we won't need him anymore." That man was one of the many who had just come to pick cotton.

CHARLES WHITFIELD

Sharecropping on Mr. Gus Diffee's place. Mr. Diffee, and Charlie Smith and Henry Orr, a bunch of them had big farms and they had sharecroppers. A lot of us would make a crop and they would furnish us with something to work with and they'd feed our horse. We would grow the crops and then gather them.

How did sharecropping work? Well, the landlord furnished you the land, a mule, and a team to work. You planted it, he furnished the seed, we would plant the corn and cotton. I done all the work. They'd plant sorghum and peanuts. A lot of 'em had hay. You'd give half of what you made. If you made two bales of cotton, your landlord got one of them. And you would divide your corn. If we didn't use the corn, we would sell it. We carried corn to the mill and had it ground into cornmeal. I would sell my cotton and pay my grocery bill.

ALBERT M. WILLIAMS

What we had to do to get along at that time. "I talk more from the standpoint of a sharecropper, which I was. Whatever we raised, we shared it with the owner of the farm. The biggest part of what he gave us was not cash money. We had a store where Mother and Daddy would go buy groceries. But mostly we lived off what we raised. That was the key.

Chapter 3

Income Sources for Buying Food, Seed, and Other Items

KATHRYN BAJOREK

Saving seed for next spring. We saved our seed from year to year for fertilizer and borrowed against the mules and harness until fall when we picked cotton and paid them back. We had an old house on our place. When my dad would take a bale of cotton into town to sell it, the gin would pay us extra money for the seed. Or we could say we wanted to save our seed. Then he would go back and load up the seed and we would unload it in that old house. That seed is what we used the next spring to plant. Finally, Dad bought a tractor so he could use the money for gas for the tractor.

R. L. "BILL" CARTER

The government put in a cannin' kitchen in Monroe. We would furnish the beef and go down there and work. They would take half of the meat and give us the other half. There was a big bunch of us kids so when we got our crops worked out, we'd work the neighbors' crops. That's how we made the money to buy our clothes. We bought our own overalls and shirts from Sears & Roebuck from the time we got big enough to pull a cotton sack.

What we had left over, we'd give half to Dad to help on the rest of the expenses.

Gettin' back to that cannin' kitchen, we also used the money we earned by working the neighbors' crops to buy our own canner and sealer. We ordered copper-coated cans from Sears & Roebuck catalogs. They had a lid that would go over that can when you would put the beef in it, then turn this crank and it would seal it on the can. They also had a thing on there that would cut the lid off. You didn't throw that can away. You just threw the lid away. The can kept gettin' shorter and shorter until it got too short to be used.

MARY LEONA CARTILLAR

Daddy would borrow enough money to buy the seed. After Roosevelt went in, they give us seed. But, he borrowed the money to put in the crop and he always bought a hundred pounds of sugar—and enough flour—I imagine that was a hundred pounds, too, of flour, but I know the sugar was a hundred pounds 'cause we had to make the sugar do all summer.

Ever once in a while we got to make chocolate candy out of it. But, most of the time we had to be real careful with the sugar to make it last. He always bought enough of beans and sugar and of course salt to do all summer. And we was always glad when it come time for the crop money, 'cause we knew we was gonna get groceries out of it and he stored them. And Mama—she was really a scrounger, too—would even go to the woods when she was able and make railroad ties. Daddy could really hew, but he had arthritis so bad I helped him saw the timber down to make ties lots and lots of times.

Making and selling railroad ties. We done it with a big old long saws with handles on each side. Daddy would say, "Don't ride the saw." I've sawed a many a tree down. When I married I would tell Russell, "I'll help you saw that tree down." And he would say, "No, you're not either! You don't need to be sawing on them trees."

There was a place along the road, like in Hickory Ridge or Tilton to sell them. When it was muddy, we'd get the ties out with a mud boat—built with boards set on runners that would be pulled out of the forest by horses. Even after I got married, Russell got them out like that. He would take them to Tilton. Daddy took his to Hickory Ridge.

Daddy took a broad axe to hew them ties out. He could really do it—Mama could, too. I never could use a chopping axe, but they could. Russell, my husband, would make eighteen to twenty a day. All the money we got from selling them went for food.

GENEVA LOUISE BECKETT COTTON COLLINS

My dad saved the best of each crop for seeds for the next year. Because of his forethought each year, our crops were bigger and better. For instance, his peanuts would have five to seven peanuts in a shell, which we broke into sections and planted in rows. We had the biggest watermelons, etc., in our part of the country. We raised more than we needed of some of our crops—corn, pork, sorghum molasses, peanuts, etc. The money from that bought what we didn't raise—sugar, coffee, salt, etc.—plus our shoes and what clothes we bought.

LETA TISDALE CURRY

Daddy lost his job after the stock-market crash. There was a sawmill near our farm. Daddy worked there in the summer between planting and harvesting seasons, also in the winter. Soon after the stock-market crash, the mill closed and daddy was without a job.

Then he started selling vegetables, eggs, butter, etc., to the grocery stores in Hot Springs. In the fall we had income from our cotton crop. Winters were hard. We had little to bring in money except for the sale of eggs, butter, and a few chickens. Daddy did trap rabbits and sell the furs. A beautiful creek ran through our farm, which was near Jack Mountain. There was one spot deep enough to swim. That was fun! After a rain, we would dig worms and go fishing. That was another source of good food.

WILLIAM EDWARD DELAMAR

We grew cotton, raised corn, peas, sorghum, apple trees, fruit trees. I had to chop a lot of cotton but we didn't have to hoe corn too much. I started picking cotton when I was six or seven years old. We raised our own meat, canned a lot of food. And you know, Mama knew so many ways to prepare food.

Take chicken, for example. You can have fried chicken, chicken and dressing. And there are so many things you can do with a chicken. First thing, they lay eggs. You can take them eggs and you can make cakes and pies. You can even make ice cream. Then you can turn around, you can take that same chicken and make dressing or you can fry it.

Now let's go back to your cow. [chuckling] You can milk that cow and have milk to drink. That same cow, you got your butter. Now you can use your milk to make cakes or you can just drink your sweet milk. Or make buttermilk. You can also kill that cow. You can have steak and hamburgers.

Now let's go back to the hog. He ain't fit for nothing until he dies. Then you can have pork, stuff like that. Otherwise, that's all he's fit for. And we had peach trees and apple trees.

S. LORETTA ECHOLS

Many of the farmers lost their land during that time. It was because they couldn't raise the crops and then the crops they did raise were not valued at very much. A lot of it had to do with the fact they couldn't afford the seed to plant it and then to pay help to work it with them.

JEAN EDWARDS

People who had credit through the banking system would borrow money for crop seed; others would sharecrop. My family mortgaged land for seed. We would have trouble paying that mortgage. Normally, you wouldn't have problems getting it extended if you were known as a person of your word. At that time, a man's word was his bond. That's all changed. A person today doesn't mind coming up to you and making you a promise even though they know when they make it that they are not going to keep it. Back then, the lending agencies respected both whites and blacks if they were found to keep their word.

GENEVA KING EMERSON

First, we saved seeds from one crop, traded with neighbors. If seeds and fertilizer had to be purchased, Dad might have to borrow the money against the next crop. A failing crop meant you might lose your milk cow or farm. Dad tried to save money from a cotton crop or produce from "truck patches," which he peddled in town for such expenses. Also, we cut cedar posts, cross ties, and stave bolts from our woods. At first, all these things had to be hauled via horses and wagon. About my mid-teen years, Dad got a pickup truck. What a labor-saver! But it also added somewhat to our labor—we began going over to the bottomlands in northeast Arkansas each fall to pick cotton between times of harvesting our own cotton, corn, and truck crops, and making sorghum molasses.

ROBERT EUBANKS

We raised cotton, but we were never on what they called "relief." There *was* a welfare program back in those days. The government would send a relief truck by and they would ask if we needed clothing, other commodity goods, but we never took anything from them because our family was proud. They were determined to make it on their own and they did! We had the hogs, we had the garden, we canned, we sold the cotton. And my dad was always a tither. When he sold his cotton, 10 percent of that went to the church. We believed that God would provide for us if we were faithful to Him in that respect and God wonderfully provided.

BEULAH LEE MCLEOD EVANS

Low prices made it hard to make any profit. Well, like we raised cotton and we had cows and sold milk and we sold other crops. But the price of everything was so low, that you didn't make anything hardly because of the drought, and then the price of it was so low that you couldn't get anything back. And people that had money in the bank went and drew their money out to where the banks—if you had money there— they ran out of money and they couldn't get loans, so if you had money in the bank a lot of times you couldn't get it out.

We sold cream in ten-gallon cans. We hauled it up to the gravel road to the mailbox, which was about a mile or a mile and a half away. The mailman would pick it up and take it to Imboden, and then it was shipped by train to Kansas City. We sold it in ten-gallon cream cans, and then the cans would come back and we would pick them up and we'd have more.

[Daddy] had to borrow money from Grandpa McLeod to make a crop on, and he had never borrowed money in his whole life before, and that broke his heart. I don't know how much he borrowed. I think it was three hundred dollars. That was a lot of money then. We used it to buy seed for cotton and corn and beans. And fertilizer. Of course, we fertilized a lot with manure from the barn.

JEFFIE NAOMI MOSER FELTON

It was hard times. I was livin' at LaCrosse when the Depression hit. We had a cow and sold milk. We just got by, that was all. Homer worked everywhere he could get a job. We had cotton one year and he worked out and me and the kids, O. C. and Imogene, picked cotton. We'd take our water to the field and we had a little ol' dog and we had a can to water it in. We'd just set it down and that dog wouldn't bother it. We'd set it down under a bush and that little dog would go get its can and take it and set it close. It would tickle us to death to see her carry that can.

[Author's note: Excerpt from "Down Memory Lane." Used with permission.]

ZELMA LOUISE HAMILTON

Saved seed from year to year. At the beginning of the Depression, my daddy put in a store, gas tank, and kerosene tank. He was wealthy at the beginning but because he had such a good heart, he let people come in and get sugar, big sacks of flour, cans of lard on credit, so his money diminished.

VIOLET HENSLEY

My daddy didn't have any money. The only cash crop he ever had was a little cotton crop and only one year before I got married, we made two bales of cotton. We had two acres and I remember him being so proud of himself. He said, "Two bales to the acre! Two bales to the acre!" We got a little money. He made fiddles to pay for a wagon. He traded a fiddle for a cow and a calf. Traded a fiddle for a shotgun. He had to buy our shoes and I guess shoes cost about fifty cents.

We grew our own broom corn, made our own brooms just like the bought ones. I can still make one of them brooms if I had the broom corn. We also made our own soap. We butchered a hog, and we had a salt box at the end of the house. We would cover the meat deep with salt to keep it from spoiling. Or we would smoke it and hang it. But that meat would get pretty rancid before we would get it eat up.

GEORGIA HEARN HILL

Daddy saved his seed from one year to the next so we would play in the cottonseed because it was soft. We played in it until it was time to plant. In the wintertime Dad worked in the log woods. He would catch a truck about four o'clock in the morning and get back home about five in the afternoon. He was one of the main sawyers that cut down those big logs. He would always tell how he could throw that tree. There was a certain way they sawed and a certain way they had to throw that tree—how it would fall. My dad was a professional in log cutting. They used horses and wagons to haul the logs out of the forest.

ROSCOE E. JEFFERSON

It was so dry back in the thirties we didn't hardly make anything. We didn't have no money! You didn't have to *have* money. We raised what we eat. We didn't even go to town. If we did go to town, we walked about six miles to the nearest town, which was Clifton. We done some trading—meat for chickens, etc.

WILLIS MAGBY

The Depression didn't end for us in 1939. I was working for six bits (seventy-five cents) a day in 1940. Things started getting better with the start of World War II in 1941. In that year, I was working in a lead mine here in Point Cedar for thirty cents an hour. It's about a mile and a half or two miles from where I live.

MAMIE H. MAYS

We saved our own seed. Somebody over in Malvern would furnish you seed and we would use our first bale of cotton to pay the debt. We couldn't go into Malvern but about twice a year in the wagon. They would put us in the wagon and take us into town about thirteen miles (about three hours) away to get a new pair of shoes.

WILLIAM WESLEY MILLER

You'd do just about anything to make a little money back in those days. We walked about a mile and a half, or two miles, to school. I would make rabbit gums, or traps. I had about six or eight of those rabbit traps out in the field and I'd bait 'em with corn, or something like that. Then if I were lucky I'd catch a rabbit in the nighttime. I'd get up early of a morning before I went to school and run my traps to see if I'd caught any rabbits. I'd get 'em and put 'em in a chicken coop 'til I got home in the afternoon. Then I'd take the entrails out and get 'em ready. You didn't have to cut the heads off or anything, you just had to take the entrails out.

If I had three or four rabbits, the company in St. Louis would send me little tags with wire that you could tie around the legs and it had the address on it. You didn't wrap 'em or anything, all you had to do was just hang them out on the mailbox on a nail, and they went through the mail like that. They'd be kinda frozen, or cold, and you'd just tie 'em together. They'd go to St. Louis and they'd sell the meat, and the hide was used to make gloves. I'd get about fifteen cents for each rabbit. It didn't cost much to ship 'em up there; in fact, they may have paid the postage. But that's how you could make a nickel or two and money was scarce. And that was fun too, especially when you had a good run and got three or four that night.

[Author's note: Excerpt from "Down Memory Lane." Used with permission.]

WILLIAM PIERCY

If Mother had extra eggs, she would sell them. She would also sell milk and cream. We raised a batch of pigs. A two-hundred-pound pig would sell for two dollars. Mother saved all the seed. We only had milk cows and beef cows. When Dad would butcher a beef, he would wait till cold winter, then eat off it before it spoiled and canned all that was left so it didn't spoil. Later, Mother canned sausage and beef. Dad sugar-cured pork, hams and bacon, smoked it with green hickory and hung it in the smokehouse and kept it locked.

UNION H. STOUDAMIRE

There would always be someone in the family somewhere that made a little money. They used to cut down these trees. I have did it a many times—pulling that cross-cut saw with my uncle. I had one end, he had the other. We cut timber out in the woods and he would cut them blocks off and stand them up and I had to peel 'em. I had to get all that bark off that tree. Then he would split it and carry it to the mill. They called 'em "bolts." They would get about three or four dollars a load, which would buy us some groceries. You could get as much for three or four dollars that you could get now for twenty-four dollars.

HERRON THOMPSON WHITFIELD

During the Depression I took a day's work for 50 cents when I could get it. And there was days I couldn't get as much as fifty cents. My wife was an orphan and she received a little bit of assets, nearly $400. We bought 50 acres of land with that, and we got started. We just worked and done without. We lived up there by Dolph 'bout seven or eight years. We first bought 50 acres of land with a little house on it, fenced, and with a spring. We paid $200 for it. The next thing that came available was 40 acres for $150. When that 40 acres come available, I'd got enough by then to have some cattle and some horses, and I was just getting started when the Depression came. I could buy a weaning calf for $8—if I could get the $8. Nellie would just support me in ever way. It couldn't have been no better.

[Author's note: Excerpt from "Down Memory Lane." Used with permission.]

FLOYE WINGFIELD

When food became scarce. That's when President Roosevelt came in with the Red Cross. They gave out commodities once or twice a week. Families were assigned a certain day. We would go to Arkadelphia to pick up flour, meal, and lard. We didn't get sugar so we used honey. Nearly everybody had a beehive or two. We had four beehives. The boys would catch them. You could hear them swarming. The boys knew how to hold the bags. If you did certain things, the hive would fall into the bag. If the queen would fall into the bag, the whole bunch would follow. You'd bring the bag home and put it in one of the hives. Ours set beside the garage.

Selling butter and eggs. Mama sold the cream from her milk. After they put in the creamery here, you could put the milk into a five-gallon can and take that to the creamery near the old Welch's Grocery Store in Arkadelphia. You could churn it and sell the butter. There was a family—they weren't sharecroppers because everybody owned their own land and didn't have enough to share. But they worked by the day and they had nothing but children. Mama gave them the skimmed milk—a gallon a day—so that the children would have milk to drink.

Chapter 4

Crops and Gardens and How They Were Worked

DOROTHY COX COSTON ASHLEY

Almost everything we ate was produced right there on that farm. The only things we bought were flour, sugar, salt, coffee, soda, and baking powder. If we could grow it in our garden or field or if it grew wild in the surrounding woodlands, we could have all of it we wanted when it was in season, but we knew we had to work for it. Mother milked the cows early every morning so we produced our own milk.

Cotton was our cash crop. We also raised what we called "field" corn, which we used on our dinner table when it was fresh and fit for human consumption. It was gathered in the fall to be used for animal food and to be shelled and taken to the mill to be ground into cornmeal. The tops were cut from the mature corn while it was still green and tied into bundles to be left to dry in the sun before it was brought into the barn to be fed to the animals in winter. Daddy planted purple hull peas between the corn rows. We always had a large garden with a variety of vegetables. We ate a lot of dried beans and peas and baked sweet potatoes during the winter months.

Peanuts were used to increase cotton yield. There was no local market for the peanuts, but the county agricultural agent convinced Daddy that his cotton would yield as much or more cotton per acre if he made his rows a bit wider and planted peanuts between the rows of cotton. After we had picked the cotton and harvested all of the peanuts we wanted for our use, Daddy turned loose the hogs he was planning to slaughter that winter in on the peanuts and fattened them in that way. We also raised cane sorghum to be made into sorghum molasses.

Daddy planted vetch and lespedeeza as winter cover crops to replace the nutrients into the soil. We planted large beds of sweet potatoes, which were our staple for our winter dining table. We planted watermelons and cantaloupes in the open fields because they needed a lot of space for their vines. These were shared freely with anyone who happened to visit us when they were in season. We ate watermelon twice a day as between-meal snacks when they were in season.

Daddy also kept honeybees—until mother developed a serious allergy to bee stings. Daddy realized that Mother was much more important than honeybees were, so he had to get rid of his beloved honeybees.

KATHRYN BAJOREK

Picked cotton to get "furnish money." We saved our seed in the fall to use the next spring. We worked behind mules. I have a copy of an agreement when Dad borrowed money against the mules, the harness, and the bridles and all of the blinds until fall when we picked cotton—to get furnish money [basically, a bank loan]—and paid it back. When my dad died, my husband and I bought part of his equipment so that it would pay part of the expense of Daddy's burial. We still have the cotton planter I rode. When I look at the equipment I think, "Lord, have mercy on all the people that had struggled so."

With the furnish money, my dad would get fertilizer. Later on when we had boll weevils really bad, he bought poison. We shelled two No. 3 tubs of corn and gave one to our neighbor for grinding our meal for the winter. If we didn't have flour we had cornmeal gravy.

BESSIE BLACKNALL

We had a potato kiln. My daddy would get a pile of logs and cross them together, and we'd put a little dirt on it. Back then, what one family had they shared it with others. We'd have so many potatoes we helped the neighbors out a whole lot. If one had peaches and the other had potatoes, for example, they would swap them out and share them.

Cutting out potato "eyes." With the Irish potatoes, they would cut out the "eyes" and that was the seed of the potato. If some of the sweet potatoes, what they called the little "runt" potatoes, were left over, they would make a potato bed. We would pile dirt on top of the sweet potato bed. As they came out, they would grow the potato slips. This is how they got their sweet potatoes.

How we kept peas. When the peas were fresh, we would shell them and eat them. But the leftover peas in the field would dry. When the peas dried, we would get them out of their hulls and put them in a milk churn or sometimes we put them in a can. There was a little thing called Hi-Lyte that came in a can and was used to keep weevils out of the peas. They would put the Hi-Lyte in the churn or wherever else it was kept and then cover it up where it couldn't get any air. This way they had dried peas to eat and seeds for the next year.

Do neighbors share today the same way they used to? No, sir! [laughing] You get it the best way you can! Today, if I get mine, I keep mine. You keep yours. We look over there and we don't care what the other person over there is doing. I would like for it to be the way it used to be but that's not the way the world is today.

LOUISE J. BRANT

We raised corn, garden vegetables, and hogs. The boys would take corn on a horse to the gristmill on Saturdays to the White Rock community where it would be ground into meal. We killed hogs but because of the size of our family, we never had enough meat to last the winter. Mother would get up and fry a chicken for breakfast.

We raised sorghum and had it ground up into molasses made in a sorghum mill just about a half a mile from where we lived. Mother made biscuits every morning and other times during the day, too, if we didn't have corn bread. We killed hogs each year, salted down the meat, put it in the smokehouse. My dad started the commercial watermelon growing in that part of the country and raised more than anyone else in the area. He discovered that the sandy loam of the creekbed on our farm was good for growing watermelons.

R. L. "BILL" CARTER

We raised cotton and corn and hay. And of course we had a garden. We'd usually plant peas when we laid the corn by. We had cows. It was open range back then from the time I was born until a few years after we moved back on the eighty acres. We branded our own cows with a big "C" on their hip, a half-crop on the right and a split in the left of the ear. Everybody's livestock run outside except the mules and horses when we was farming.

DOYLE L. COLLINS

We had a garden, chickens, hogs, and cattle, etc. That provided vegetables and meat. We also utilized our orchards to a great extent and mother nature provided us with berries and grapes. Also small patches of cotton, corn, hay, wheat, and oats. Fresh meat in summer and fall was provided by hunting squirrels, rabbits, and by killing chickens. There were no deer. We tried cotton and hogs for cash crops. Mostly raised our food and preserved it.

Gee! and Haw! We would use mules and horses for plowing. Double shovels would require just one horse or mule. Cultivators required two horses or mules to straddle the rows. We would walk behind the cultivators, holding the plow close to the cotton or corn, hollering, "Gee!" and "Haw!" depending on which way we wanted the horses or mules to go. If we wanted them to go to the right we would holler, "Gee!" to get closer to the row. If we wanted them to go left, away from the row, we would holler, "Haw!"

GENEVA LOUISE BECKETT COTTON COLLINS

My family raised almost all our food. Our crops were varied. We had a large garden, with cabbage, lettuce, beans, peas, tomatoes, onions, radishes, okra, and turnips. We stored the turnips over the winter in a mounded hill. Daddy would pile up the turnips, cover them with hay, then cover that with dirt. When we wanted turnips, we would dig into the side of the hill, get what we needed, then cover the hole. They never froze or went bad!

Our farm was wonderful—lots of corn, peanuts, and a large watermelon and cantaloupe patch. Everything that would grow in Arkansas, we raised. We only bought things in town like sugar, coffee, flour, salt, and pepper. Everything else we raised, selling what we didn't eat or can. We ground our own meal by taking corn to the gristmill in Paris.

My dad and brothers were the ones who tilled the farm, mowed and baled the hay. We raised corn, peanuts, sugar cane, so our cows and horses ate well. Dad killed hogs in the early winter so we ate very well, too.

S. LORETTA ECHOLS

Even in my earliest recollections of things, times were really hard. I very well remember basically what we raised on the farm and in the gardens and we had our farm animals. That was it. We raised food to feed the family rather than as a cash crop. It was probably 1939 or 1940 before I ever bought a loaf of sliced bread. Usually by the time [you fed your family] you had doctor bills you'd try to pay off.

GENEVA KING EMERSON

Folks bartered a lot. For example, Dad would gladly work for any neighbor within walking distance who could afford to hire a worker for even half a day in exchange for return help or produce from the neighbor's farm which we could use. Sometimes he could get enough work to earn a piglet or two to fatten for meat or increase his own livestock. My mother and the neighbor ladies exchanged help on canning days. They saved printed feed sacks and swapped with each other to get enough to make dresses.

Neighbors swapped plants and seed, sometimes planning ahead so that each one started more tomato, cabbage, collard, or sweet potato plants than would be needed for that farm, then exchanged plants with each other. A neighbor was always welcome to extra strawberry, raspberry, or currant plants and seedling peaches, plums, and cherries, or surplus plants previously mentioned. When there was a need, folks divided what they had, whether or not they had a surplus. And days of hard labor were often given, even as a farmer's own crops were in need of harvesting, when some neighbor was seriously ill or injured.

ROBERT EUBANKS

Dad did the plowing and was very particular about how close those scrapers would go into the cotton row. He could run those scrapers within an inch or so of the plants, and scrape away the Bermuda grass. Then we would come along with the hoe and clean out the rest of the grass. My dad prided himself on having a weed-free cotton patch.

We didn't have lightning rods, which were big ol' clumps of Careless weed, Johnson grass, or something like that stuck up above the cotton. Dad would not allow this in his fields. "Careless weed" was a big ol' red weed. The hogs just loved it. We would go out in the cotton patch, cut Careless weeds, bring them in, and feed them to the hogs.

We would kill rabbits and eat off the land. We did a lot of that. We had swamp rabbits back in those days that were half again as big as the cottontails. Every once in a while, Dad would bring in a 'possom. We didn't like 'possom all that well, but when it was fixed properly, it was edible. Had a lot of fat on it that you could parboil and cut off. The lean meat on the 'possom was very good. We didn't have many 'coons. It was mostly rabbits and squirrels and in the summertime we would have frogs and fish, too.

MARY EATHEL FREEMAN

There was no fear of going hungry. North and back of the house was a fertile vegetable garden. Another small vegetable garden was at the foot of the hill next to the road. I'll never forget how large and sweet the cabbage heads grew that spring in 1936. Mama could cook cabbage that was delicious. We made sauerkraut in half-gallon jars. We canned green beans, peas, corn, pickles, chow-chow (relish), and every kind of fruit we could gather. We really felt blessed with plenty of food.

MARY FRANCES LOVELL IZARD

Chicken droppings for fertilizer. Although our chickens ran loose all day, at night they roosted in the chicken house and we shut the door to protect them from wild animals. After a year, the chicken house floor was deep with chicken droppings. Our first chore then was to scrape all the droppings out of the chicken house and scatter them over the garden for fertilizer.

Sweet potato slips. They were grown by planting several sweet potatoes in a large tub. Only about one-half of the potato was under the soil. The potato would soon begin to grow new plants which were pulled off (called "slips") and planted in the field.

We canned what we didn't eat. Surplus fresh vegetables, such as green beans, fresh black-eyed peas, and butterbeans, were canned. The potatoes were dug and stored and the onions were pulled and hung to dry. The cabbage was made into kraut and cucumbers into pickles. Around the last of August or the first of September, the fall garden was planted. The fall potatoes were very important because potatoes were a staple in our diet. Daddy always planted a large patch of turnip greens. We ate the young greens first and later when the turnips (sometimes called the "roots") began to mature they were added to the greens.

Gathering wild fruits and berries. Since our house was located where the road ended and we were completely surrounded with woods, there were plenty of wild fruits and berries we could gather. As soon as spring came, my dad began to look for patches of wild blackberries and wild huckleberries. My mother made the best berry cobbler in the whole world. I have tried to duplicate it but I cannot.

Preserving turnips for winter food. As winter approached there was a danger the ground might freeze and kill the turnips. In order to save the turnips for winter food, they were pulled, the tops were cut off, and the turnips were placed in a large hole dug in the garden and covered with dirt. This process would preserve the turnips throughout the winter. If we needed turnips to cook, we just dug some up and replaced the dirt to preserve those that remained.

Peas and butterbeans were allowed to mature and dry in the field. When they were sufficiently dry, they were picked and stored for winter food. The peanut plants were pulled when the peanuts matured and then stored in the loft of the barn to dry. After they were dry, we picked them off the vines and had roasted peanuts. There were also pumpkins and sorghum cane.

ROSCOE E. JEFFERSON

We grew cotton and corn. We also grew cane to make molasses. We raised watermelons. Loaded 'em up, went to Cotter, and peddled them for thirty cents to forty cents each. We would take a beef in a wagon and peddled it. People would walk up and tell us what they wanted and how much they would give.

Mules were used to work the crops. With the double-shovel plows we had then, we would plow one side of the row and come back on the other side. Back then, that's what we grew—cotton and corn. We had about forty acres, which was all you needed with one team. Each one of us four brothers—Guy, Faye, Shelby, my brother-in-law, and I—that farmed together would have a hundred acres of corn. We'd average around ninety bushels of corn to the acre. After we got our tractors, we grew lots of corn.

KENNETH GUY LACY

Picking and canning blackberries. My mother always wanted to can fifty half-gallon jars of blackberries every summer. My dad and all the children who were old enough would take two buckets—the older ones would have gallon buckets, the younger ones would have smaller ones they could carry. We would go out in the woods to pick wild berries and would pick until we had enough to fill all fifty jars. I guess we had blackberry cobbler about every week of the year. It took a lot of food to feed nine people through the winter.

EVELYN LANGLEY

We sold cream once a week. They'd test it [for richness] and you'd get paid according to how much it tests. We'd have a little money out of that, you know, to go on our groceries or buy a yard or two of material. We churned all the time. We made our own butter and buttermilk. I love buttermilk. We raised everything we ate except the flour. The little store kept it in a barrel. The storekeeper weighed it out in a paper bag. It costs about ten cents a pound, I guess. Then later on, it got to coming in those sacks that we made clothes out of.

County furnishes a pressure cooker. The first thing that came into the community that helped a lot was a woman who was kind of knowledgeable about all things. The county furnished her with a pressure cooker and she'd go from place to place and help the women can their stuff. And that was really a lot of help because we didn't have a pressure cooker. And my mama would have been afraid of it if she had had one. But finally, she got one and this woman would help her can, shell peas all day long. All the food had to be pressured or it would spoil. We had to dry our peas and cook 'em dried.

DULAS L. MASSEY

We grew cotton, corn, hay, and sometimes we had a few chickens and eggs. We had a concrete storm cellar and that's where we stored our food. It was probably ten feet by ten feet. In our garden, we raised onions, turnips, green beans, peas, and my uncle would give us a pig every year. We fed the pig slop. When hog killin' time came around in the fall, we would kill our pig and that would be our meat. Sometimes the pigs became pets but we had to eat.

Baked sweet potatoes? When my brother was about four years old, he was playing with matches and burned our barn down. The barn crib was full of our sweet potato crop, so all our friends had baked sweet potatoes for a while.

MAMIE H. MAYS

We had to prepare the garden each year. We would turn the ground over, smooth it down, and then use our horse Teddy Bear to pull a plow to make the row for the plants. Mama put up stakes at each end of the garden and ran a string across to the other end and we used that as a guide to make the rows straight. Arkansas has a lot of rocks, and it was my job to pick them up, put them in a bucket, and carry them to the rock pile. Then we used a hoe and rake to plant and to keep the garden weeded. Water was hand-carried from our hand-dug well. That was a big job digging that well, but we had good fresh water.

No mules for us. "Teddy Bear," a big, black work horse, was a good horse and understood us most all the time. We didn't ride him but used him for work. Most of the farmers we knew had at least one work horse. He loved for me to feed him carrots and other vegetables from the garden. My father chewed Union Standard tobacco and Teddy Bear would get into my father's shirt pocket to get his tobacco. Dad only gave him a very small piece. I kept asking my daddy why didn't Teddy Bear spit the tobacco? "Well," he replied, "I guess animals just don't spit."

When Teddy Bear was sold. "When we sold our farm and moved back to Kansas, we sold Teddy, two cows, pigs, chickens, laying hens, and my dog. To me, they were like part of the family."

Cotton was our cash crop. We'd get maybe seven cents a pound when the hard times come. We just didn't get much for what we sold. We also sold corn. We had a country grocery store—a "peddler," we called him—come by once a week. He'd buy the eggs. Whatever we had to sell to him, he'd buy it from us and trade it out. We had a vegetable garden—potatoes, peas, beans, and onions, beets. Anything we could grow, we grew it. Daddy was a good gardener. We also had chickens, ducks, and geese. We always had geese in the cotton patch to keep the grass out, you know. Every farmer had a bunch of geese in his cotton field. They would pick it clean. Sorghum we used to make molasses.

LLOYD A. PERRY

Different kinds of plows used to prepare crops. We used mule power to work with shovel plows, cutters, or sweeps—there were various sweeps for different types of plowing: buzzard wing, solid sweeps, the heel sweep that cut lots of weeds. We used calf-tongued cultivators. Sometimes, a longer type of tongue (narrow) plow was set in front of a larger sweep to open the furrow ahead and pull the larger plow deeper into the soil. We liked the John Deere walking cultivators and International tools if we could get them.

To keep our mules in working shape, they needed at least about a half-dozen good ears of corn and a little fodder each day. That was a problem during droughts. If corn was scarce, our chickens might be lucky to get the seed heads of sorghum. Besides our corn fodder, we saved peanut vines for hay, and the cattle relished them.

Stave bolts—An extra source of income for farmers. In the early thirties, stave bolts for barrels began to be cut in this country, so farmers with woodland found a new form of extra income by working at that on the side. Bolts had to be thirty-eight inches, made from the first or second cuts from a white-oak tree. There could not be one flaw in the wood. A cord of stave bolts measured eleven by four feet instead of the eight feet by four feet for a cord of wood. A cord of stave bolts brought eight to eleven dollars.

Sometimes we cut "quarter bolts." That is, we split the bolt into quarter sections. Done right, those brought ninety cents. That's compared to picking cotton for fifty cents, eventually one dollar per one hundred pounds. Hill cotton had to be awfully good for a good cotton-picking hired hand to get two hundred pounds a day.

December and January were the fur-season months. We trapped the wild animals that would destroy our crops and farm animals, skinned them, stretched and dried the hides for sale. Cold winter weather was needed to produce the quality fur. People in the big cities wanted those fur coats. I remember furs bringing about thirty-five to fifty cents a piece, up to five dollars, depending on type and quality. That was a lot of money back then. At the Batesville market, I heard of one nice 'coon hide that brought twenty dollars! We didn't use to have raccoons around here, but they're thick now. Much later on, in 1980, fox and coyote hides were bringing twenty dollars. But there's no market for fur now, and there's so many wild animals, farmers can't raise anything because of them anymore. We didn't waste the animals we killed. We ate the 'coons, 'possums, squirrels, and rabbits, and were glad to have the wild meat.

WILLIAM PIERCY

Drying and canning food. Potatoes, tomatoes, cabbage, pinto beans, butter beans, green beans, black-eyed peas, onions, sweet corn, sweet peas, popcorn, Swiss chard, spinach, and mustard greens. We dried or canned or stored food for a full year out of that garden. My mother would can one thousand quarts and put them in the cellar. The smokehouse and cellar were locked during the Depression because people were starving and would steal food.

UNION H. STOUDAMIRE

Everything that farmers planted we had it then. We made our own syrup, raised our own corn, all kinds of peas and beans. And gardens! We had everything in that garden that you have in a plantation today. We raised our own food, our own meat, we killed our hogs in the fall of the year and would hang them up in the smokehouse. They made smoke on them and they cured them hams. Oh, Lord! If people were living like that today, there wouldn't be near that many people needing counseling and having different kinds of complaints!

EVA SMOKE WELLS

One of the things about my love for gardening. I'd get out working with Daddy someplace. We had a big tomato crop over here. I'd be working with him and he would tell me that he had this dream that someday he would manage to raise enough [money on] farming that he would be a little better off than he was. He would tell me how if we would plant a half an acre of tomatoes and take care of them, that we could make so much money off it. Well, that depended on weather and whether we got the right seed and got it worked and whether we could sell. Maybe someone else would have some they were going to sell too!

ALBERT M. WILLIAMS

Cotton was our cash crop. We raised cotton, corn, peas, peanuts, potatoes, sorghum for sorghum molasses. Around 1918 or 1919, we got a fairly good price with cotton. That was war time, you know. Cotton went up to twenty-three or twenty-four cents a pound. But when the Depression hit, the price went down to four cents a pound. And that was ginned cotton. A bale ran upward from four dollars and fifty cents a bale.

We used two kinds of plows. A middle-buster and a shovel plow. First thing we would do would be to lift up the land with the shovel plow. Then we would come up and open it up with a middle-buster. Most of the time, we would use two horses. Sometimes we would have a single mule or horse to pull the plow, but most of the time we had two. We worked from dawn to dark.

FLOYE WINGFIELD

Income/expense estimates based on hope. You counted so many acres of cotton and averaged out what you hoped it would make each acre and you planted your acres to go with what your expenses would be that year so you'd have enough money to run the year on. We also raised corn and beans.

Getting corn ready to be taken to the gristmill. You carried your own corn to the gristmill to have cornmeal made. I remember my father always shelled off the two bottom rows of the toughest corn down where the shuck comes on. He would shell that off in the animal bucket and then he would break off the top end where the worms might have gotten in and eat. And then the center you took to the gristmill. You ate the best part of it. And the other went into the feed for the animals. None of it went to waste. Sometimes the animals were hungry enough to eat the cobs, too.

JACK WOOD

The sweet potato house. The sweet potatoes fed us the year around because Daddy had what we called a sweet potato house that had a lot of bins inside it where he would put potatoes to dry them. He'd have a little round stove in there. He would put kerosene in the bottom. And that thing would keep the temperature at seventy degrees all the time, morning and night the year around. We'd have sweet potatoes the year around. He called them kiln dried. They would taste just as good after a year as they were when they were harvested. We would produce about six hundred bushels every year.

Two-day round-trip by wagon to Hot Springs. We'd go up one day and come back the next. I had a mattress so I would just sleep in the wagon that night. Back in those days, you didn't have to be afraid of anybody bothering you.

We kept two dogs to keep snakes away and to alert Mother about hoboes coming to the house. We lived not far from the railroad tracks and in those times there were many riding the trains. They would get off at the switch near our farm and mooch for food. I never knew Mother to turn them down. She wouldn't allow them in our home but would have them wait outside until she could carry food out. The dogs kept watch on them while they waited for the food.

[Author's note: From an interview by Maggie King, his granddaughter.]

Chapter 5

Typical Chores around the Farm

DOROTHY COX COSTON ASHLEY

I helped gather fruits and vegetables and prepare them for cooking. If it had to be shelled or washed or whatever had to be done to it before it could be cooked, I helped do it. We had vegetables from the garden and plums, blackberries, and huckleberries (small blueberries) that grew in the surrounding woodlands. Feeding the chickens, gathering the eggs, and bringing in wood for the wood-burning cookstove and fireplace were some of my chores.

I washed canning jars when mother was canning. I churned milk to make butter. I made my own bed and hand-washed and dried dishes almost daily after meals. I helped hang clothes on the clothes line in the yard and helped take them down and bring them into the house after the sun had dried them.

At a very early age—I really don't know at what age—I learned to thin cotton with a hoe. I also learned how to hoe the grass from the cotton and corn in the spring. Traditionally, most of the girls I knew did the same thing. I was taught to pick cotton at a very young age and I always had a quota of cotton to pick each fall. All of the girls I knew picked cotton. I helped cut and save corn tops, tied them into a bundle and left them sitting in the fields until they dried. Then we fed them to the animals in the winter. We didn't produce any hay but we did produce corn tops.

We had cows, hogs, and chickens. Arkansas had an open stock law at that time so the milk cows were turned loose to graze in the nearby woodlands during the day, but their calves were kept in the barnyard to make sure that the milk cows would return to the barn every night. The calves were allowed to nurse both morning and night, but Mother only took milk for our use in the mornings.

The cows were separated from their calves in the barnyard at night after the calve nursed so there would be plenty of milk the next morning for the calves and for our personal use, too. Female calves were kept to increase the size of the herd, but male calves were sold in the fall, which was another source of income.

BESSIE BLACKNALL

We would sweep the yard and clean up the house. We'd use a hoe to dig up pieces of grass that was growing in the yard. We'd scrape it clean. We'd make brush brooms out of dogwood branches and tie 'em together with a piece of wire or a string. For light, we used a coal-oil lamp. We also helped in the garden and in the field.

We raised our own hogs and chickens. The way we cleaned out the chickens before we killed them was to put them in a coop for two weeks and feed them corn to fatten them rather than let them eat whatever they found in the chicken yard. To clean out the hogs, we would take them out of the pens and put them on a wooden floor for two weeks and feed them with a special food before they were killed rather than letting them run wild and eat whatever they could. By the time we kept up the hogs for two weeks, they were cleaned out. The hogs and chickens must of thought they were *living* in heaven but didn't know they were about to die and *go* to heaven! [laughter]

HERMAN BLOCK

We did it all. Chopping and picking cotton, cutting wood, planting the garden, harvesting crops, feeding farm animals, bringing in water, making fires, etc. Both my sister and I fed chickens, cows, hogs, etc. I cut wood. Went to pond to get wash water. Sister gathered kindling. We both chopped and picked cotton. Mom and Dad chopped and picked, too. I helped my dad plow fields, etc. Mom made buttermilk. When we got an icebox, we bought block ice and kept it in the icebox.

LOUISE J. BRANT

Pulling "hot caps" off watermelon plants. The boys did the things that helped my father and we girls didn't do very much except we carried drinking water to the fields and fought the sand burrs to get there. We laid watermelon vines because you had the rows far enough apart that the trucks could get down to the middle. Sometimes the vines grew over into the roadway between the rows of watermelons and we had to pull 'em back. As the plants began to emerge, we tore open the wax-paper "hot caps" that had been placed over the soil to warm it a week before planting. This helped the watermelons to germinate more quickly.

Getting up on cold mornings to milk the cows. I didn't dread it because I was so intent on going to school. That to me was an obsession. I would sacrifice anything to do that. My younger sister liked to dress really well. If she got a job on the farm where she could make some money to buy a new dress, she would take off school and do that. She used to say that when she got up in the mornings and started saying she was going to quit school, that I would start throwing clothes at her and yelling, "Get up! Get up!" When she graduated from high school, she wrote me the nicest note about how I had pushed her through school.

R. L. "BILL" CARTER

The girls helped hoe and pick the cotton. The boys gathered the corn and all the hay with an old pitchfork. The girls helped milk sometimes. To cut the hay, we had a ridin' mule-drawn mowing machine. I got the old hay-rake wheels out here [by the house] now. And I got the wheels off an old cultivator here. And I got an old one-horse turning plow sittin' down there next to the fence. Using a cross-cut saw, the boys cut the wood to cook and heat with. The girls would help wash the clothes on a washboard.

I plowed my first mule when I was eight years old. They put me runnin' the middle with an old mule that was twenty years old. He knew what to do. I had to hold my hands up like this [demonstrating] to hold the plow. The plow had a heel-sweep on it and then a plow right in the center. When I would get down to the end, the old mule would want to eat a bite or two and he'd graze a little bit. Then he would turn around and I'd drag that plow around. We had to saw our firewood. I started sawin' wood when I was about eight years old.

DOYLE L. COLLINS

Our milk cows, horses, mules, etc., were named and became vital family members. Money was hard to come by. To illustrate just how hard it was, I'll relate a little story: We owed a local merchant a good-sized bill and had no money for payment. In 1937, mules became expensive because of President Roosevelt placing tenant farmers on thirty to sixty acres and letting them farm the land to help pay it off. It was a law of supply and demand so far as the prices of the mules was concerned, because they became vital to those small farmers for cultivating their cotton and other crops. The result was that they became high priced. My dad gave "Old Beck," one of our mules, to the merchant to cover the debt. The family cried and so did the merchant.

S. LORETTA ECHOLS

I was the water-bucket carrier. I recall that we all needed to help and one of the things I was saying to someone the other day is that one of my earliest remembrances even at three or four years old was carrying a bucket of water. We didn't have electricity down in the country where we were. And we didn't have the pumps and wells for our homes. Now there was a few of the farmers out toward town that did have wells but that was not our situation out where we were. We had to carry the water to wash the clothes from Bayou Meto, a long and deep waterway, and everyone had a pail or a bucket.

BEULAH LEE MCLEOD EVANS

I was a tomboy. When they was planting cotton, I would ride on the log. It was a big log. The planter would go ahead on these rows and then there was a big old log. And a horse tied in front of it and would pull that log. If I thought they were not looking, I would ride that log. The log was used to cover the crops. It would log over where they had planted. I would tell the team to stop and go and turn at the end of the row. At the end of the row, if a horse turned too sharp, the log would turn over. That was dangerous, you know. I'd have to run to get out of the way. I never could make a straight row. But I could drive a horse and make a straight row because that horse was going straight.

MARY EATHEL FREEMAN

In those days, there was "men's" work and "women's" work. My brothers helped cut wood and took care of the horses. My sister and I carried water. It was quite a chore. Some came from a well about one hundred yards away. Some came from a spring down a steep hill from the house. Every afternoon, we filled stone jars, buckets, kettles, and pans for night and morning use. That meant several gallons and several trips. There were also trips during the day when water was needed for cooking and canning.

Hunting for guinea nests. I liked to help pile and burn brush, to see a cleared field. The timber was cleared away from the house place. The poles that were left made a good place for the guinea hens to hide their nests. It also was a good place to protect the little ones from hawks. It was fun to hunt for the guinea nests—sort of a super Easter egg hunt. Lela and I hunted. We listened for a special sound the female made after laying an egg then tried to spot where she was. Later, we could come back and go directly to the nest. [But] we never went directly to it while she was close by. Guineas were especially clever about protecting their nests. They would stop using it if they found it disturbed. We learned to never take two eggs at once. I actually believed that they could count to five. At the age of eight, I must have been an odd child.

GEORGIA HEARN HILL

Daddy had a way of hollering that caused the cows to come running. They always knew his holler. We would "feed up" before night time. We would feed the cows, hogs, chickens, gather the eggs. Go get the cows and bring them up in the afternoon and have them ready to milk in the morning. We had to bring in the night wood and that was about it.

After we got big enough, we chopped cotton. We thinned out the corn, thinned out the sorghum for molasses. We just pitched in to do whatever needed to be done. These were our chores. Getting up on a cold winter morning before dawn to milk three cows was a chore we had to do. After we milked the cows, we brought the milk back into the house.

KENNETH GUY LACY

Mom did the milking, usually having only one cow, sometimes we had two. After straining the milk, she would let the cream rise to the top, skim it off and churn it for butter. She usually did the churning and used a big crock jar with a wooden dash. She would also take care of the chickens, setting hens and gathering the eggs. I guess she didn't trust any of us to help.

Dad took care of feeding the cows, horses, and "sloppin'" the hogs. We boys helped Dad when we were old enough. He cut the wood for use in the fireplace in the winter and the fire wood for the cookstove. My brother and I had to carry in the wood, and helped split the wood for the cookstove when we got old enough to handle the axe. The girls helped Mom with the housework.

DULAS L. MASSEY

I was responsible for all milking and most of the weeding. My chore was to milk the cow, work in the garden, feed the pig. Go get the cow in the afternoon and drive her in from the pasture. I had to milk the cow. I also churned the butter. One time we didn't have very much to eat, but we ate a lot of molasses, butter, corn bread, and buttermilk.

MAMIE H. MAYS

Everybody had something to do. One thing I did was to always go with my mama to milk so I could hold the cow's tail while Mama milked to keep it from slapping her in the face. I learned to milk before I was nine years old. Getting out on cold winter mornings was hard to do but you didn't know any better. That's how we were raised. We had to bring in the cook wood and the kindlin' used to start the fire.

RUPERT E. "BUSTER" MELTON

I was the milk man. I never was around sisters because my grandma took 'em. There was just me and my brother Marvin left. I had another brother but he died as a baby. So it was just me and him and my dad. We went along with that until me and my brother got old enough to get out on our own farm.

We followed an old mule. We didn't have none of this easy-working stuff that they got now. We didn't know nuthin' about it. We got out there with an old cotton hoe. We had it pretty rough. See, my mother died of pneumonia when I was seven years old. You get that pneumonia back then and it would shut your wind off. A lot of people died. That one year, somewhere around 1917, it was so bad. Everybody said it was the "flu." But when it came through that year, man, it cleaned 'em up. People just died, died, died. If it hadn't been for that, they would probably have lived to be old. My mother died young.

EVELYN M. COONFIELD METCALF

I gathered eggs, fed the chickens and pigs, and cared for my dog Pooch. I also got the cows up, helped on wash days—usually Mondays—and helped clean up the "two-holer" outdoor toilets. We would use the laundry water to clean the two-holers. We put lye in the holes to help keep them clean and smell good. Outside of that, keeping up with the house, working in the fields whenever needed. There were just the three of us—Mom, Dad, and me. Dad was busy on the farm doing all the things we had to do to keep things going.

WILLIAM PIERCY

At the age of six, Bill started going to the barn every morning to milk fifteen cows by hand. He also was assigned to take care of their one sow as well as cows and chickens. To breed the sow, Bill's father would tie a rope to one of the sow's front legs and take her to a neighboring farm. With no bull, the family had to transport cows for breeding. [*Violet Piercy, quoting Bill.*]

JOANNE RIFE

I had to clean and fill the kerosene lamps each day. The hired hands carried in the wood for heating to the wood box on the back porch. Mama made lye soap from fat drippings and Merry War lye, and I had to shave it into finer pieces for her to dissolve in the Monday wash water. The women took care of the chickens then, before the county agents told the farmers that chickens would replace the nutrients in the soil and provide a cash crop. So Mama went to close up the broiler house at night where she raised those fryers that she made fried chicken from all summer, three meals a day.

UNION H. STOUDAMIRE

I had to wash the dishes, milk the cows, and churn milk. I also had to feed the chickens. I tried to pick cotton. Some of my friends would be pickin' four hundred and five hundred pounds of cotton a day and I couldn't do it. We got about a dollar twenty-five or maybe two dollars a hundred, something like that. I went many a day and couldn't make a dollar. We milked whenever we got up in the morning. Somebody had to! It never got too cold for us to go out and milk those cows to get the milk and butter.

EVA SMOKE WELLS

I was Daddy's boy. Being the oldest, I was the one who was taken out to the field to work and I loved it. My sister next to me, and she's dead now, preferred to stay in the house with mother and grandma. She didn't like fieldwork at all. My oldest brother was sickly as a baby. He almost had what they called "rickets" back in those days, a lack of certain proteins and things. He didn't drink milk. That was the problem. He just would not. But they caught it in time and he is all right. He lives back over here now [pointing] between my house and where my parents lived.

Getting up to milk on cold winter mornings. Sure, it was tough, but you didn't think about it for the simple reason that was the way you were brought up. It was your job and there was nobody else to do it. And if you wanted milk to drink, you had better get out there and do it.

LUCILLE RIDER WILSON

Memories of picking cotton. I remember one day in August when the nights began to be chilly. I knew that soon cotton-picking time would come. I dreaded it, the soft wooly worms on the leaves, the wet cotton which shrunk my fingers uncomfortably. I could pick two hundred and fifty pounds. We had long sacks that dragged on the ground. I just weighed ninety pounds but I'm a fast worker. I also had to hoe the weeds out of the cotton and corn in the spring. Today you see corn growing and it's real crowded. We thinned it to about a yard apart, I guess because the ground was so poor. We picked up what we cut down and fed it to the hogs.

Chapter 6

Preserving Milk, Making Butter and Buttermilk

KATHRYN BAJOREK
One of my very first jobs that I was assigned was to dig a hole in the ground. My dad gave me a stick and said, "You dig this hole. It doesn't matter if it takes you all day. Dig the hole, then stick this stick down in there. If the stick goes all the way down, then the hole is deep enough." And then we had scraps of burlap that Dad had cut out of sacks so that I could put them into the hole and press them against the wall of the hole. After we would set a gallon of milk in there, we put burlap over it. Then we had some old rags we had cut out of cotton sacks that had tar on the bottom. We would put this on the top so that the dirt didn't sift down on the milk. Then you would put the dirt on top of that and the milk would stay cold.

BESSIE BLACKNALL
The black folks would put a bucket of milk in a container of water to keep it cool. The white folks would tie a rope to the bucket and drop it down to just above the water in the dug well. We had about three milk cows so whatever milk we got in the morning was gone by the end of the day. To make ice cream, we would wait for an ice truck to come through. There was a hole in the floor of the smokehouse. Mama would wrap the ice in an old quilt and put it in the ground so we could go out and take a chip out of it. The ice wouldn't last very long but long enough to make some ice cream.

[Author's note: Mrs. Blacknall is African American.]

R. L. "BILL" CARTER
We had a hole in the ground in front of the water pump to put the milk and butter in. The hole was about five feet by five feet and five feet deep. We had milk but we didn't know how to make milk gravy. In fact, we had never heard of milk gravy until my oldest brother went to a CCC camp around 1939 or '40, where he was a cook. When he come home, he showed Mama how to make milk gravy. They took him right out of the CCC camp and put him in the army.

GENEVA KING EMERSON

During hot weather, we quickly strained the fresh milk. This was to remove any foreign particles that might have gotten in it into lard or syrup buckets washed with soda and water and air dried, then hurried it down to our spring branch where it was set into the cool water with a rock on the bucket lid to keep it from turning over. (By the way, those buckets were scarce, highly prized, and often bartered for.)

Dad built an enclosure of rails around our spring and a portion of the branch to help keep out stray dogs, etc. During cold weather, souring milk was no problem—back then milk froze setting right in our kitchen. My grandparents and great-grandparents had dug cisterns in which they hung by rope or chain a bucket in which they placed the milk container. A few people had spring houses and a great-great-uncle had a cellar built over a spring.

ROBERT EUBANKS

The water table in Dyess was very high. You could dig three feet down like a posthole and in thirty minutes that hole would be full of water. You could take a ten-foot section of a two-inch pipe, put a sand point on it (sharpened point to keep out dirt), drive it down with a pipe driver until about four feet were sticking out.

Then you would take the cap off the pipe, screw the pitcher pump on to it. You could prime it by pouring water in it and almost immediately you couldn't pump it dry. If you kept pumping it, the water would get real cold. We would take this cold water, set it in a foot tub (big enough to put both feet in), put the bucket of milk down in it, and stir the milk. The cold water would take the heat away from the bucket of milk. Do that three or four times and eventually you'll cool the milk down to a drinkable temperature.

BEULAH LEE MCLEOD EVANS

Turning cream into butter. We had a big old stone churn and it had a lid with a hole in it and a paddle and you would churn that up and down until the cream turned to butter. We separated the milk with a big separator. I never have figured out how they worked, but you poured it in this big old bowl and you turned on the separator and cream would come out at one spot and milk without the cream would come out at the other. Well, we drank that milk that was separated. We didn't drink the whole milk. We drank that other. We drank that like water.

How we kept the milk cold. We didn't have an icebox to keep it in. We kept [the milk] in a big concrete cooler Dad made in the well house. We drawed water to put in that concrete container. It was cold well water that come from I don't know how many feet down because we lived up on a hill. That was a deep well and that water was cold.

MARY FRANCES LOVELL IZARD

How we strained the milk. After the milk was brought to the house my mother strained it through a clean cloth and poured it into gallon jugs that had secure lids. A stout cord was securely tied around the neck of the jar and the milk was lowered into the well. At meal time someone would pull a jar out of the well for us to drink.

Soured milk churned into butter. As soon as the milk soured, Mama poured it into a churn to use for making butter. The milk usually had three or four inches of cream that had risen to the top of the jar. Even when the milk was good, the cream was carefully removed from the milk and added to the churn. The remaining milk was still plenty rich to be considered whole milk. We would not drink what is today called two percent or skimmed milk. That was fed to the pigs (called "Blue John" today).

Making buttermilk. When my mama considered the milk ready, she churned it until the milk fat turned to butter. She removed the butter from the churn and pressed it in a wooden mold. The buttermilk was saved for drinking or for cooking. Any excess was fed to the pigs.

WILLIS MAGBY

We had an icebox later on but when I was a kid, we had a dug well. We would let the milk down in a water bucket and let it down to a certain place where we would tie a string on the rope to stop it and tie it off and let that bucket hang over the water which would be about sixty degrees, which was a lot cooler than it was in the hot summer. In the wintertime, sometimes your milk would freeze if you just set it out. But it never froze if we put it down in the well.

Chapter 7

Hog Killing: A Special Time of Year

BESSIE BLACKNALL
How we prepared the hogs. We'd cut [the hogs] up into different sections of meat—hams, the shoulder, the tenderloin. The chitlins you can still buy now but they are expensive.

We took the hog's head and made homemade sauce, sometimes called "hogshead cheese." To preserve the meat, we put it in a smokehouse. We'd cut an opening in the floor, dig a hole under it, then hang the meat up above it on a piece of wire. We'd make a fire in the hole underneath the floor but smother the fire and just let it smoke. The flies wouldn't bother the meat on account of that smoke. The smoke would cure the meat, which is why they could keep it for such a long time and keep it from spoiling.

ROBERT EUBANKS
We would butcher as many as four hogs at a time. Those hogs would weigh about two hundred pounds a piece dressed out. We cured our own meat, we had a smokehouse, and we would salt the meat down for the two-week period, wash the salt off it, then put a tub in the smokehouse, build a little fire up in the tub, and smoke the meat. Later on, we got real modern and started putting liquid smoke on it to give the meat a smoky flavor.

We made our sausage by the tub full. Mother would cook the sausage, make it out in patties. We seasoned it ourselves. She would make the patties out and fry 'em in a big ol' skillet. And then she would stuff them in fruit jars and pour the grease in on top of 'em. And I'll tell you, friend, that you have never had good sausage unless you've had *that* kind of sausage. She was canning vegetables, green beans, butter beans, and any kind of beans we could have, she would can them. We also canned fish.

BEULAH LEE MCLEOD EVANS
When we killed hogs we hung them on a high pole and scraped the hair off and gutted them. Then we laid them on a long platform and cut them up. We stripped the fat from the intestines and from the other cuts and rendered it

in a large wash kettle. We salted down the middlings [what we call what you make bacon out of now] and hung the hams and shoulders and smoked them by putting ashes in tubs then slow burned hickory wood for days. We ground the meat for sausage and fried it and put in jars and poured grease over them and sealed them and turned the jars upside down so the lard when cool resealed them so no air could get into them. We would divide the liver, heart, back bones, and ribs with neighbors who helped with hog killings.

Salting down the middlings. We had a smokehouse and when we would kill hogs, they would salt down the "middlings" in a great old big container—and just cover them up with loads of coarse salt. It had to be where they wouldn't ruin, you know, and cool them like that. The hams and shoulders we would hang up and had the big No. 3 tubs, and have ashes in the bottom of them and hickory. You would make coals and have hickory on top of that and smoke them hams and shoulders. You smoked them so long and then you could hang them up in the smokehouse where they'd keep a long time at a time.

MARY FRANCES LOVELL IZARD

Hog-killing day. We always had two pigs that Daddy would get as soon as they were old enough to leave their mother. Male pigs were cheap because the farmer only needed one. Our pigs were happy pigs. All they did was eat, scratch, and roll in the mud. After a simple surgical procedure performed by my father while a friend held the pig still, we were looking forward to some fine pork meat. They were fed corn, pumpkins, milk, table scrapes, and any vegetable or fruit that was unfit for the table.

These pigs didn't know it but like Hansel and Gretel, they were being fattened for a reason. One cold winter day they realized what it was. Daddy watched the weather carefully. If the pigs were slaughtered too soon, the meat might spoil. When he decided winter was here to stay, he notified the friends and relatives who were going to help and the next day the work began. Children were sent to the house and not allowed to watch.

After the pigs were killed they had to be dipped in a large barrel of scalding water and then their hair was scraped off. After this was completed they were ready to be butchered. Mama was sent large pans of fat that was cut from their abdomen (known to us as sow belly). She already had a fire under the wash pot and she cut the fat in pieces and cooked it in the pot until the clear liquid fat came to the top. As it rose to the top she dipped it off and put it in five-gallon metal buckets for storage. All afternoon she cooked the fat and put the liquid in the cans. After the fat cooled, it became solid again and was the lard we used to cook with. The usual yield was about five cans full and was sufficient to last until the next year.

Daddy had built a large box in the smokehouse. He had also bought several large sacks of salt. After the pigs were cut up, everything but the pieces he intended to grind for sausage were packed with salt in the box in the smokehouse. The salt acted as a preservative for the meat. The sausage was ground and seasonings mixed in and stored in a cool place.

If there was more than we could eat before it spoiled, a few days later Mama would cook it, pack it in fruit jars, and can it. The liver was a special treat for hog-killing day. It was sent straight to Mama, who sliced it up and cooked it for lunch for us and also for the ones who came to help. Fresh liver, brown gravy, and hot biscuits is a meal that is hard to beat. Don't confuse our liver with what you can buy today in the supermarket.

My dad loved the brains. Mama cooked them mixed with scrambled eggs. He didn't have to worry about sharing. None of us would touch them. The cured pork furnished seasoning for vegetables. It could also be sliced and fried. The lard was used in cooking and also as a necessary ingredient for making soap. Also one of the big hams was saved for Christmas.

RUPERT E. "BUSTER" MELTON

My dad never did buy no meat. There wasn't no stock law back then, so you could have all the cattle you wanted. And my dad had that bottom full of hogs and cattle. 'Course, we never did kill no beef. He'd kill a hog ever time he got a notion and we got out of meat. We'd go out and get a hog, shoot him between the eyes. We didn't kill hogs unless it was cold weather. We had a smokehouse. We put the meat away in that, but our fingers and hands would be as cold as the dickens.

EVELYN M. COONFIELD METCALF

Butchering time was in October. We had one pig and one calf for us. This could take three or four days. After the cow and/or pig was killed, it had to hang and "bleed out" for a couple of days and then it was dipped into a large vessel of boiling water before it was butchered. We canned some of the meat, made head cheese and made sausage patties and laid them in lard to keep. We sugar-cured and smoked the meat and it was hung in our screened-in back porch.

UNION H. STOUDAMIRE

How did we kill our hogs? We knocked them in the head! [chuckling] We had goats and we had hogs, we had cows, we had chickens, we had ducks and turkeys. One day we went to fishing and I had about a dozen young turkeys. We had this big shaggy dog. We came back to find that dog had killed all them turkeys. Some of them wasn't dead when we got out there, lying out there kickin'. I was a-hollerin' and cryin' and I told my husband to "Kill that dog, kill that dog!" But when he went to get the gun to kill that dog, I started crying, "Don't kill my dog!" We were so crazy about that dog. He was shaggy and pretty with that long curly hair. My husband said, "I should turn around and shoot you!"

JACK WOOD

We butchered about six hogs for food each year that would weigh about two hundred pounds each. That made a lot of pork. We would smoke the hams over a hickory wood fire then place a sack over them and slice portions off as needed. The rest of the hogs were buried under salt in a large vat and portions were removed from the vat as needed. We also kept about twenty-five chickens that kept us well stocked in eggs the year around. In fact, Mother would trade the old country peddler eggs for things like salt, pepper, aspirins, and other small items as needed.

> [Author's note: From an interview by Maggie King, his granddaughter.]

Chapter 8

Making Molasses, Peanut Butter, and Such

R. L. "BILL" CARTER
When molasses turned into sugar. We had a sorghum mill at our house when we lived at Rich. My great-uncle Joe, my grandfather's brother, put it up there. People would haul their sorghum there and burned some mule power to grind it. That old mule would go around and around. I remember one time we had our sorghum cooked just a little too much and it turned to sugar. We had a fifty-five-gallon barrel we set out in the yard, put those molasses in there and they all turned to sugar. We put a cover over the barrel and nailed it down. We'd take a tin plate, dip some out of the barrel, and put it on a cookstove. When it warmed up, it would be just as good and fresh as it was when it was first made.

GENEVA LOUISE BECKETT COTTON COLLINS
We had our own sorghum mill. Daddy would build the fire under one long metal pan that was divided into three different holding compartments. Our horse was hitched to a long pole that was attached in the middle so that the horse went around in a circle which mashed the sugar cane to get the cane juice out, which in turn went into the first compartment of the cooking vat. There it cooked until my dad knew it was time to pull the plug and let that juice into the second compartment in the cooking vat.

After that juice went into the second compartment, Daddy would send more sugar cane through the press, filling the first compartment again. When the first juice finished cooking in the second compartment and Daddy knew it was time, he would let that juice into the third compartment to finish cooking to become sorghum molasses.

This process started early in the morning and was a two to three all-day event. The first part, of course, was taking the wagon down the rows of sugar cane while some of the men cut the cane and put it in the wagon.

BEULAH LEE MCLEOD EVANS
We made sorghum molasses using a horse or mule. The mill was over here

[pointing], and you fed the cane into a machine and the machine was run by that mule going around and around pulling this long stick like they do in sorghum mills, you know. And then the molasses would run down.

There was a great big furnace and he had to keep that really burning, you know. There was a tray that was about this wide and about eight feet long, and that juice would go into that. You would cook it until it got just right, and Daddy always knew just exactly when to take that off and run it into gallon buckets. Boy, he could make good sorghum! And a lot of times my brother would haul sorghum to the County Farm. It was like a nursing home where old people went. He'd give it to them, but he'd sell it to neighbors. And he'd even sell it to stores. We always kept enough for ourselves.

Daddy would make peanut butter. And we would pull them peanuts and stack them up in an old house that we had over on Grandpa McLeod's place. When there wasn't anybody living there, we would stack them real high and let them dry out. That place had a fireplace in it. In real cold weather we would go over there, sit right down in that floor and pick off peanuts. To make peanut butter, Daddy would first roast the peanuts, and then we'd shell them out. They had to be all good; he wouldn't put a bad peanut in there. He'd grind them over and over again in the sausage grinder until they was right creamy, and then he'd put olive oil in them and salt. He made good peanut butter!

MARY FRANCES LOVELL IZARD

Making molasses. The sorghum cane was cut and hauled to someone in the community who owned a sorghum mill. At the mill the cane was run through a press to extract the juice then boiled to the right consistency in a large copper pan heated with a wood fire. The resulting sorghum syrup was put in one-gallon metal buckets and stored for winter use. The sorghum was processed by the owner of the sorghum mill for a share of the syrup.

ROSCOE E. JEFFERSON

We made lots of molasses and we used horses to pull the mill. They just went in a circle. You had to grind the juice out of the cane, then you cooked the juice. We had a tray fifteen feet long and four-foot wide with dividers in it. We would pour the juice in one end of the pan and as it cooked, we moved it down to the other end, where we would fill buckets. We would make sixty or seventy gallon of molasses a day. It took six or seven gallon of juice to make one gallon of molasses.

EVELYN LANGLEY

Another recipe for making peanut butter. We'd shell our peanuts and husk them and roast them and grind them in the sausage mill and make our own peanut butter. She would take this biscuit dough at breakfast and roll it out thin and make our crackers. She'd roll it up then and put it in the oven and make it brown. She'd put that peanut butter on them homemade crackers and we'd have crackers and peanut butter. Kids nowadays would just faint and fall over, I guess, if they had to do what we had to do. But see, we didn't know any better. Everybody done it. So everybody was just in the same shape.

Chapter 9

Meeting and Taking Care of Hoboes

R. L. "BILL" CARTER

Hoboes riding the rails. When we left down at Rich, there was a grove of trees between our house and the Missouri-Pacific railroad tracks, which went through Monroe into Brinkley. It had a passenger train and it had a freight train. There were always hoboes on that freight train. They would stop by our house occasionally to beg for something to eat.

I remember one time one of 'em wanted to stay all night in the hay barn where we put up our hay, but Dad wouldn't let him stay there, afraid he would burn the barn down. So there was a big pile of leaves out in the yard and he asked if he could stay in that bunch of leaves. We let him stay in there but when we got up the next morning, he was gone.

BEULAH LEE MCLEOD EVANS

We had hoboes that would come by the farm. I don't know what they were doing that far off the road, but they found it and knew that people along there treated them. They would stop, and they would want to work. A lot of them would want to work for food. I knew Mother wouldn't have work for them. I remember one time her letting them split some wood. But she'd just always feed them. And I remember especially one. He was starved to death. We canned sausage then in quart fruit jars and would turn them upside down when we were ready to eat them. He come in and she fixed a whole quart of them sausages and a big pan of biscuits and gravy, and he ate every bit of that. That man was hungry.

LaVERNE WILLIAMS FEASTER

Many times there would be hoboes on the trains. Nobody got off at the Dixie Plantation because only white people were riding the train. Black people couldn't. Only time they rode the train was when they were maids and could ride with the white folks.

[Author's note: Mrs. Feaster is an African American.]

ROBERT FLANNIGAN SR. • S. LORETTA ECHOLS (Joint interview)

FLANNIGAN: My mother always had some food because [hoboes] were coming to our back door every day, and she always had something for them.

ECHOLS: We were frightened sometimes [by hoboes] if my dad wasn't home during the day. If people were just walking down the road, and you didn't know them, they were probably hoboes. We didn't know them. There was a railroad that went through, probably just two or three miles from the areas where we lived and she'd say, "Now when they get here, I want ya'll to stay back from the door and I'll see what they need." Although she would say, "No, they won't hurt us," we were just cautious. Out in the country by yourself, you didn't let them in, but they sat on the porch and they ate. Whatever you had, you shared it with them. You didn't say, "I don't have anything." If the family had anything, then that hobo would have something, too.

SUZANNE GROSS MARKS

We lived about three blocks from the railroad. Lots of tramps rode on the train boxcars. I think back they must have had our house marked some way because we had lots of tramps coming by for food. One day one came (they all worked for food) and wanted something to eat.

We had a black man [named Charlie Holland] working for Dad—with the cows, plowing the garden, etc., helping around the house. He was going to Philander Smith in Little Rock and Mother asked Charlie what he had fixed the tramp to eat. He told Mom [he had fixed] a sweet potato sandwich. I suppose we had baked sweet potatoes the day before. So Charlie just made him a sandwich from those. We thought it was a funny sandwich but I guess [it was] healthy and filling.

Chapter 10

Making Our Own Clothes, Going Barefooted

DOROTHY COX COSTON ASHLEY
We got a new pair of shoes each fall after school had started. We usually got a pair of white shoes some time in the spring that we wore to school in the fall during the first few weeks of school. Daddy would often apply a half-sole to our new fall shoes before we wore them so they would be sure to last us the full year.

Some of the fabrics for our clothing were ordered from Sears Roebuck catalog and some them came from little dresses that adults had outgrown or didn't want anymore that were handed to Mother. And some were made from printed fabric that flour came in. Most of our underwear was made from bleached fertilizer sacks. But she made all of it by hand until she got that sewing machine.

BILLIE BAGBY
Mama made me underwear—bloomers—out of flour sacks. I also remember Dad would resole our shoes by putting them over a "shoe last"—shaped like a foot and upside down. He would cut pieces of leather and attach them to the shoes with tacks. Sometimes you would feel those tacks inside your shoes. But I always got a new pair of high-top shoes and new stockings and some long underwear when school started. I hated that long underwear! Many times, I would get outside of the house, pull it off, and stick it up my underwear.

KATHRYN BAJOREK
My mother made everything we wore from feed sacks. She could look at someone and just cut a dress to fit them. That's how later on when people would come to the house, they would bring my mother a piece of material or another sack or two for her to cut out the dress so they could make their daughters a dress. Mother could just look at them and ask, "Do you want a collar? Do you want sleeves? Do you want pockets? Do you want trim?"

We treasured twine from the feed sacks because we girls had long hair and

Mother would French braid it. She would dye the twine in poke salat berries. She used hulls from walnuts and hickory nuts. By boiling these on top of the stove in a tub, she could get different colors. She would get a batch of twine and then crochet that and make like a lace and would zig-zag it into our French braids to decorate our hair. Everybody thought we had really done something. And then she would crochet lace out of those different colors of twine from the feed sacks. Also, she would take the top of a feed sack, cut the tar bottom off, and Daddy would use it when he was crawling under the car if the windows didn't work.

She also made bloomers to match with the buttons on the side to hold them up. We had these "tokens" issued by the government that Dad would take into town to get some sugar, flour and other food items. Since there was no elastic, she would make us bloomers that were kinda like a skirt. They would have a top around them with a button on the side.

Part of our job was to take old shirts that people had given us or thrown away. My dad wore long-handled underwear, so after it wore out we cut those up and used them for dish rags, and sanitary pads because they were soft. We saved the buttons. My mother would make bloomers for other people that didn't know how. Every button was treasured because there was no elastic anyplace.

R. L. "BILL" CARTER

We worked in the cotton fields to buy our clothes. We got fifty cents a hundred pounds. When we picked two hundred pounds, we would get paid a dollar. We would make a dollar a day to chop cotton ten hours a day. Mama made our shirts. Back in them days, the girls wore slips and bloomers. That's a far cry from today. Mama used flour sacks to make dresses for the girls.

How Dad measured us for new shoes. When we lived down at Rich and got the last bale of cotton out, Dad would put a piece of paper down on the floor, we would put our foot on it and draw around it and he would get us a new pair of shoes. They had to last all winter long. A lot of times the soles would come off 'em and we'd have to put 'em back on with wire. When it got warm weather [chuckling softly], we went barefooted.

MARY LEONA CARTILLAR

Flour sacks used for making dresses and sheets. We would buy a twenty-five-pound sack of flour and it was in print. We would get the same print and I would make dresses out of it. Now that was after I was married. But, we used "shorts" sacks, too. I didn't ever have a sheet but what was made out of short sacks. "Shorts" is a feed for animals. It come in a white sack, it was a hundred pounds. After I married, all my sheets were made out of short sacks. I never had a bought sheet. It's pure cotton.

Old habits hard to quit. I still don't throw nothing away. I can tell you, you can't hardly get through my house. And the food—this is one of the lessons I learned from the Depression—I got it stacked in cans. There is no earthly way I can use it! When I find a good buy, I buy lots of it and it's just setting there. I've got freezers full. I look at it sometimes and I say, "Lord, I'll never eat all that stuff!"

What about shoes? Well, mostly, we went bare-footed. I can tell you. I don't suppose we ever had over one pair of shoes a year, till our feet outgrew them.

WILLIAM EDWARD DELAMAR

Shoes wrapped in "croker sacks" in winter. When it was sleeting and snowing, folks would would wrap their shoes in "croker sacks"—burlap bags—which feed used to come in. Today, you see 'em used sometimes to sack Irish potatoes. Regular shoes would slip and slide. But the croker sacks were just like a ground-grippin' tire on an automobile. A lot of them old folks would take the croker sacks, dye 'em and make clothes for girls. Or shirts. I have a sewing machine in our house that my mama operated with her feet. I was ten or eleven when Mother brought it to the house.

ROBERT EUBANKS

We didn't have a whole lot of money so clothing was a little difficult to come by. We ordered it from Sears Roebuck. This was exciting. Once a year we would order our clothes to outfit the family, which included seven kids. When the box came, you couldn't believe how big it was.

We got one pair of shoes a year. I had narrow feet so I never had a pair of shoes that fit me until I was twenty-five years old. While I was in the Seminary, I started selling shoes for Sears & Roebuck. I learned to measure feet. We had a pair of shoes in the store that was my size and when I put them on it, they felt so good that it was like I had died and gone to heaven. In my first twenty-five years, I had never had a pair of shoes that fit me.

MARY FRANCES LOVELL IZARD

We were saved by the Sears & Robuck catalog. Every fall before school started we made an order. Everyone got two pieces of material to make dresses, my brother got overalls, and everyone got a pair of shoes. Since we were children with growing feet, Mama took a piece of paper and drew around our feet and Sears used the drawing to send the right-sized shoes.

My oldest sister could sew very well. Mama had an old treadle sewing machine. As soon as the order arrived my sister began making our dresses. It was such a thrill to get a new dress. She also made our panties out of flour sacks. What we used as elastic [elastic was rationed because of the war] for the panties came from an old inner tube from the tire of a car. She cut the inner tube in strips and used the rubber like elastic. It is true that we made dresses out of feed sacks. The problem with that is we hardly ever got two with the same design.

EVELYN LANGLEY

Mama made all our clothes except the shoes. She even made our underwear. To make our clothes, she used flour sacks, mail sacks, and feed sacks that came twelve to a bag back in those times. We had a little Mr. Krafton's store by the New Home Church house and the material was just ten cents a yard. She had all these chickens and she'd sell four or five dollars' worth of eggs a week. And maybe one week, she would get enough material—us girls was little— she'd make a dress out of just two yards, which was just twenty cents. Then above what she had to have for groceries, maybe each week she'd get a dress a week for one of the children—one for a child one week, one for another child the next week, you know.

SUZANNE GROSS MARKS

Most clothes were homemade. Mother had a lady to spend two weeks and made dresses, gowns and underclothes. We knitted sweaters, blankets. We didn't get new clothes but hand-me-downs or made-overs, like maybe converting a shirt to a skirt, A lady found out how to make a skirt form a pair of pants by ripping apart all the seams, pressing them out and then turning them up the other way to make a nice skirt.

DULAS L. MASSEY

We bought our clothes with the money we got picking cotton. We boys had two pair of overalls and one pair of shoes for the winter and sometimes in the summer we would get a pair of slippers. They cost about two dollars and fifty cents a pair. I remember buying my first pair. It took me about two months to pay for them. I'd pay for them a nickel or a quarter a time. Some grown people who went barefooted all summer would come into town on Saturdays.

MAMIE H. MAYS

We had underwear made from sugar sacks on which was printed "Sweet and Pure." When we bent over, the printing was visible. That was a joke during the Great Depression. We also used flour and feed sacks to make quilt tops and hand towels. Later on, I learned to sew and made clothes for all my kids. Both of them died in their sixties. My daughter's son lives next door.

EVELYN M. COONFIELD METCALF

How my dad resoled our shoes. Dad had a pipe stand with a shoe plate in three different sizes. He would put my shoe on the shoe plate, use skin from a cow or pig if we didn't have leather, and glued the new sole over the shoes. After a few hours of drying time, he cut around the edge of the hide and shaped it to the shoe. He then used a small, wood tack around the edge. If the old sole was real thin, he glued a little piece inside—he didn't take the old sole off. When the shoes got too small for me, they were given to someone else who could wear them.

LLOYD A. PERRY

I learned how to make shoestrings from squirrel hide. It makes the best, long-lasting shoestrings. We preferred the red squirrel for the larger skin. To tan the hide, it can be buried in ashes until the hair slips, then worked until soft. But someone told me the tanning could be done by rubbing the squirrel brains thinly on over the raw side of the hide, then warming it. The hair begins to slip off right away. Then the hide should be kept around where it can be picked up often and kneaded between the hands. After a few days, it'll finally get nice and soft, then it can be cut into string.

UNION H. STOUDAMIRE

Our shoes were purchased, but [had been previously] used. Most of what we wore was homemade. We didn't have no money to buy no clothes back then. We made clothes for the girls out of pretty feed sacks. I wore many a dress to church and to school made out of them feed sacks and they were pretty. They were starched and ironed and they were beautiful. For the boys, the clothes that weren't give to them were store bought. Except that there were some cut-off pants made from flour sacks.

JAMES A. THOMPSON

The kids always went barefooted. You could walk on a rock road and it wouldn't have hurt 'em at all. Me and one of the neighbor boys were out in the yard one day and somebody broke some glass and it was covered by leaves. I stepped on it barefooted, and boy, made a big cut on my foot. I still have the scar from it. When she heard us call her, she come out. After looking at the cut, she said she was going inside to get some kerosene. That's all she had. She poured in all over my food. Oh, man, that hurt! Then she brought out an old scrap piece of cloth and wrapped my foot up in it. I went home and will never forget that.

ARWILDA WHITESIDE
We used bleached flour sacks, fertilizer bags, rice bags, feed bags, and any other sacks we got then tie-dyed them. Everyone sewed. We were always clean and neat. We wore shoes every Sunday. Our feet would be cracked and hurting. We greased them with tallow (beef fat). We had to clean the grease off before bed or it would get in the bed. We soaked and washed them in the washpan.

LUCILLE RIDER WILSON
My mother made everything—dresses, underwear, shirts. She didn't make coats, but she made everything else we wore. She was a good seamstress. She could cut her own patterns. She used to go to the grocery store at Peter Pender. The merchant up there knew Mama sewed and he would bring out these feed sacks that he had saved in matching designs so that she could have enough to make a dress. When she sewed, she had a treadle machine. I can remember there was always one of us hanging over the machine watching her sew.

In 2003, I sought to capture these moments in "Mama's Love Notes"

> Mama whistled as she sewed, the foot pedal marking time.
> Blue checkered cotton metamorphing: Now a ready-to-wear shirt for Coy,
> A love note.

> Mama whistled as she scrubbed, sloshing soapy water, stoking the coals.
> Reddened hands scrub patched overalls on the rub board.
> Reba's blue print dress drying on the fence.
> A love note.

> Mama whistled as she walked, dropping potato-eyes along the furrow
> Growing vegetables from sweat, soil and manure,

Seeing heaped up bowls of food for five giggling kids waiting on the bench,
For love notes.

Mama whistled as she poured a stream of boiling water onto chipped white plates
barely drying between meals.
Wiped her hands on her apron to start another
Family love note.

Now, Mama whistles through empty rooms echoing quarrels, laughter and tears.
Pencil and tablet in shaky hands, recounting the daily happenings around the farm.
Reaching out across state lines with
Five love notes.

FLOYE WINGFIELD

You bought a shoe that was long enough that you could wear it till it wore out, which was never more than a year. But you always had to have enough room at the end of that shoe toe that you could wear them that long. On good years, you got a pair of Sunday shoes, too. I liked my Sunday shoes. I really like shoes, but I didn't have many.

One time my brothers didn't have any shirts to wear. They had worn them out. We had to make a crop. The fertilizer came in a sack that had the picture of a Quapaw Indian on the back of it. Mama saved those sacks and washed them several times in boiling water in that wash pot. She got 'em good and sanitary then set to a sewing machine somebody had given to her. And she could sew well. So she made these shirts and she was very careful to center the Quapaw Indian right in the middle of the shirt so it would really look so good when you would see that old mule going down the row and one of the boys hanging on to the plow handles. Mama made a crop anyway and we did have a little money left after that first crop.

Children ridiculed clothes donated by the Red Cross. Some of the children wore clothes to school that the Red Cross would give them. That's the only charitable organization I remember us having at that time. The clothes they had was adult sizes that had been donated. But the children wore the adult clothes to school and everybody made fun of them,

Well, I was in the group that made fun of them. I realized later that the only thing wrong with me was jealousy. I would have just loved to have had some of those grown-up clothes on. Some of them even had high heels. Man, couldn't I have cut a path with that! [laughing]

I thank my Lord for this: There's people who wound up in California I saw maybe thirty years ago and talked to them. I told them if they knew I was in the group that had made fun of 'em, "I just want you to know that the only thing that was wrong with me was that I was jealous of what you had that I didn't have. And I'm sorry for what I did. It wasn't to be mean. It was mean but the mean part was me, not you. You hadn't done anything. I was the one who got greedy. I would have just loved to have some of those clothes you was wearing to see how I would look."

I'm very grateful I had a chance to do that. We can't always make amends for things we do that was bad a hundred years ago.

Chapter 11

How We Washed Our Clothes

DOROTHY COX COSTON ASHLEY

Mother had two No. 2 washtubs and a great big black pot. The first thing she did of a morning when she was going to wash, she filled that black pot full of water, built a fire under it, and put some soap chips in there. The first clothes she washed were the white clothes. They were into one of the washtubs and she scrubbed them on that rubboard with her bare hands. Next, they were taken and put into the wash pot and left to boil while she scrubbed the colored clothes on the rubboard.

Then the colored clothes were taken from the tub and placed in the next tub that had rinse water in it. After they were rinsed, the clothes that had been boiling in the black wash pot were lifted out with a stick like a broom handle, brought back and placed in the cold water so she could handle them. They were rinsed in two tubs of water and then placed on the metal clothes line out in the yard, attached to the lines with clothespins for the sun to dry.

It was backbreaking work and especially in the wintertime when your hands were cold and you would make a mis-lick and hit your hands on that rubboard. It was a bad job.

GENEVA LOUISE BECKETT COTTON COLLINS

On wash day, all four of us females would be out in the back yard from start to finish. My mother would make lye soap from tallow that my dad saved from butchering hogs. Then she added lye and other ingredients. My dad would build a fire under the wash pot, and we would carry water from the spring on our property to fill the pots. The pot would be brought to a boil, for the white clothes, and usually it would be my mom, Grandma, my sisters Lavada and Vera, and me washing on the rubboard and rinsing. The clean clothes were wrung out by hand until Daddy bought a wringer, which was a great help. It was hard on fingers, though, as I caught mine in it once. The clothes were hung to dry in the sun and then ironed by flat irons heated on a woodstove.

WILLIAM EDWARD DELAMAR

How to make lye soap. You can get a piece of oak or hickory—any kind of hard wood—and burn it down to ashes. Get them ashes, put them in a bucket, drill a little hole in the bottom, pour water in at the top with something under the bottom of the bucket to catch the water, and this will make lye you can use to wash your clothes. See? A lot of the things we did back then, nobody now knows anything about it.

GENEVA KING EMERSON

Washing took a whole day each week. We usually started the day before, carrying lots of water up the hill from the spring, filled a big kettle set over a wood fire and the rinse tub on the wash bench. Wash-day morning, the tub for soaking and washing was filled with hot water from the kettle. As the water was used and had to be replaced, we toted up more. Clothes were sorted, and the white or light-colored ones washed first, then boiled in the kettle. Then the colored ones were washed. Clothing with really stubborn dirt that didn't come out after the usual soaking and then being scrubbed on the rubboard with lye soap was boiled in soapy water in the kettle, then rubbed and rinsed again. Then the laundry was hung up on the clothesline to dry. In cold weather, it freeze dried.

MARY EATHEL FREEMAN

The real test [for carrying water] came on laundry day. Twenty gallons of water was needed to fill a pot used to boil white clothes. A tub of water was needed for rubbing the clothes on a rubboard using soap, and two tubs of water were needed for rinsing away the soap. The first tub was for rinsing away most of the soap; the second contained bluing for whitening and brightening everything. If it rained, water was always caught under the eaves. Tubs, vessels, and a rain barrel were placed under the drip. Almost everyone had a rain barrel.

In the summer, we did the laundry down by the creek. Water was heated in the pot for rubbing the clothes and for boiling the white things. When the clothes were hung on the line and dried, they smelled so good. All vessels were left handy for the next washing day. It was almost heaven not to carry so much water. You can see why I fell in love with the farm. I had not been so content in years.

[Author's note: Excerpt from Eathel Freeman, *Journey of a Depression Kid,* 3rd printing (self-published, 2005), 14, 19. Used with permission.]

EVELYN LANGLEY

How we made lye soap. We'd save all the meat grease we had, and cracklin's, things like that, when we killed hogs. We'd put it in this wash pot, pour in water and build a fire around it, get all the grease out of it, and buy some lye to put into it. We'd stir it with a big ol' wooden paddle till it got thick and hard. And then you'd put out all the fire under it and just let it set in that kettle until it cools. Then you'd take a butcher knife and cut it in big squares.

Chapter 12

Family Bathing Routines

CELIA ALMYRA GRAHAM ACREY

We didn't have a bathtub so we just used big No. 3 tubs. We had to bathe pretty often because we had to work in the fields. We'd come in, draw us up one of them big tubs full of water, and set it out for it to heat. And then when it would come time to come in, the first person there would take a bath. The water was never changed so everybody would hurry in to be the first to take a bath.

KATHRYN BAJOREK

We bathed once a week. We had a No. 2 washtub. My daddy saved a bucket he had bought some grease in. He used all the grease to grease the plows and the wagons, he cleaned all that out to make what we called a "foot" tub.

My mother had rags, tow sacks, old clothes that she had cut up. We would use the rags to wash our faces, under our arms, wash our private parts. When we got through with that, we would step over into that little foot tub and wash our feet. When it was warm enough, my mother and dad would take us to Bayou Meto. He would hold the boat close to the bank while Mother would turn us on our backs to wash our hair using lye soap that was made from boiling hog fat and lye in a washtub. After drying the mixture, mother would cut it into bars.

R. L. "BILL" CARTER

When we got in from the field, we had to take a bath in a No. 3 washtub. We took a bath once a week whether we needed it or not. Our outhouse was for the girls to use, and a Sears & Roebuck catalog was the toilet paper. The boys had to go to the barn or go to the woods and use whatever they could find. People need to know things like that but probably won't believe it.

GENEVA KING EMERSON

We had to carry every drop of water we used except for the rainwater we caught in any vessel we could find. We prized the rainwater for its "soft" qualities for washing, etc. The spring water was "hard," but of a wonderful taste, due to mineral content. In summer, we left water setting in tubs to heat in the sun; in winter, we heated big kettles on the woodstove or fireplace. When Mom started in to wash the flock of youngun's, everyone's face was washed, maybe some extra dirty places cleansed with a separate rag. Then the baby was washed first and right up the list by age—making me last! I wasn't very old before I decided I could do my own bathing, carrying, and heating clean water. We could bathe in a very small amount of water when necessary.

In fact, we could really conserve water! We might first do laundry, then use the dirtiest water for mixing hog feed or watering plants. The cleaner rinse water was used for bathing, scrubbing floors, then for plants. Sometimes water passed through three or four uses before reaching the plants. I've wondered why all that lye soap didn't kill them!

How often did you bathe? Main baths were usually once a week, especially in winter, but Mom carried a vengeance for visible dirt. A wet washrag was always handy for dirty faces and baby hands. We older children dipped from saved wash water or from the carrying bucket, pouring the water into a washbasin, washing our hands and faces, dumping the used water on Mom's flowerbeds by the porch, or into the hog-feed bucket in the wintertime, then drying on a feed-sack towel. In hot, sticky weather, especially when we'd worked in the fields or woods all day, we yearned for a quick dip in the branch or a wipe down with wet rags, which we washed and reused.

ROBERT EUBANKS

We had no running water, just a pitcher pump. And as many of us as there was, we couldn't take a bath at the house. So every day, when we got through working in the fields, we'd go down to the drainage canal—we called it the big ditch—behind the house. The water was always clean. We'd take a bar of soap with us and bathe in that canal, then walk back to the house.

In the wintertime, we would heat water in the old black water kettle. We had a bathtub in the house but we had no water to go in it. We would take the water out of the hot water in the kettle and pour it into the bathtub. Two or three of us would bathe in the same water. When it got thick enough you could walk on it, we'd change the water. Mostly, we bathed just once a week. The rest of the time we'd just take sponge baths, which we would do nearly every night.

LaVERNE WILLIAMS FEASTER
We heated water on the kitchen stove and the living-room heater for baths. We bathed on Saturday nights. We pumped water in tin tubs, which were put outside in warm weather. The girls bathed first, then the boys.

VERNON L. GLENN
We'd take a bath about once a week. After working, we had a little smokehouse and we would get enough water in the tub where you could have some privacy and all five of us kids would take a bath in that same water. Not all at the same time. The girls would get to go first. We had a little joke about that—that the last person that took a bath would go down in quicksand. We had a lot of little jokes that kept us going.

WILLIS MAGBY
Many times, we waited until the evening to go to the creek, which was about a half mile away. When it got too cold to go to the creek, we did without. People back then didn't take a bath every day like we do now! We'd wash off every day in a washtub. That's just the way it was. No one around here had a bathroom, so you just didn't notice people not taking baths.

Did you change the water after each person finished? Oh, no! Whoever was last had to use the water as all those who came before him. It's somethin' that none of us really want to talk about. That was some of the worst part of the Depression. What I mean is, there were times when we had to save water. A lot of people's dug wells went dry in the summertime. Or if they didn't go dry they'd get low, you know, and you'd have to conserve water. If my hands were not very dirty when I finished washing them, I wouldn't pour out the water so my brother could wash his. Sometimes we had to haul water from the river. We didn't get electricity here until 1948.

MAMIE H. MAYS
We bathed in the biggest washtub they had, a No. 3. In my daddy's first family, there were five. There were five of us in the family and we all took a bath in the same water. But we had to wash the dirt off our feet before we got into the tub. If we didn't take a bath, we just took sponge baths to wash up.

JACK WOOD
We had a cistern where we caught rainwater. In the day when we were at work or school, Mother would fill a large, long zinc tub and let it sit on a small

porch so the afternoon sun would warm the water. Then when it came dark, each one of us would get our towel and we would line up in a small room next to the porch with the youngest one first. Then the one called would step in the tub and Mother would take her large washcloth and give them a good washing. Then they would take their towel, dry off, and return to their room.

Chapter 13

Did We Know We Were Poor?

CELIA ALMYRA GRAHAM ACREY

No, we didn't realize it. We just stayed at home and worked. We didn't have nothing much but we were just happy-go-lucky. When my mother died, she left a little baby. Papa had me and my sister and some older ones, and he didn't know what he was going to do. So he asked my aunt Edna Odom, my mother's sister, if she would keep the baby for a while until Papa got back on his feet and realized what he could do. She didn't adopt him. She just kept him.

She would come to our house and we would go up there and everybody was just as happy as a lark. Papa later married a woman in Searcy that had two kids—a girl and a boy. All of them are dead now. There are only two of us Grahams still living—me and my brother. We're the only two left.

KATHRYN BAJOREK

No, not really. After I got into the seventh grade—that's when we had to start wearing shoes—my mother never went to town, and my dad—I guess it was when we got "furnish" money—he took it and bought all of us one pair of socks. The boys' socks had stripes that went around the foot and elastic around the top. We wore those every day and my mother washed them out and hung them on a wire that was connected to the stovepipe.

BILLIE BAGBY

I knew we didn't have any money, but I also knew that no one else had any either. So I never felt poor. To this day, I pick up a pin off the floor, I save it. I save rubber bands, paper clips. I did save paper sacks but I don't any more. I find it very difficult to get rid of anything. Because I might need it or somebody else may.

FRED NORMAN BLANKENSHIP

I know it now but I didn't think much about it then. There was no money because times were pretty hard. We always had plenty to eat and plenty to wear even in the bad times. We had hogs, cows, chickens, cotton, and corn.

You know back then cotton was the main crop. My daddy sold it for five dollars a bale. A lot of people would give the seed for the ginning but my daddy brought the cottonseed home to feed to the cattle. He never put any cotton into some kind of government program. He sold what he grew.

LOUISE J. BRANT

Well, we really didn't know we were impoverished because everybody else was in the same situation. And I sure hope we don't get back to those days! We just did the best we could with what we had. From the time we were big enough, we took odd jobs picking strawberries. We used to walk about two and a half miles to a strawberry patch. You'd only get one row, so you'd probably make a quarter. That was our spending money. We bought our clothes with what we had earned.

R. L. "BILL" CARTER

Everybody in the community was poor. We used to go out in the fields after frost and pull grass to put into a mattress cover to sleep on. Some folks used corn shucks to put in there. Back in them days, there were some kind of bugs they called chinch bugs. They was pretty good-sized bugs and they would eat you up at night. My mother, she'd pour coal oil all around the edges of the bed to try to keep them off us but they got on us anyway. They didn't get rid of them until they started spraying houses with DDT.

I remember one time after the crops were laid by, Dad went and got a job peeling the bark off poles. The next day he got up, and Mama said, "I don't have a thing in the world to fix for your dinner today." He said, "That's all right. I'm not going." And she said, "Why not?" He said, "I've worked as hard as I've ever worked in my life and just made seventy-five cents. If we starve, we'll starve rested."

DOYLE L. COLLINS

Yes, we knew we were poor. The reason we knew we were poor was that we had a debt hanging over our heads. To pay for a gall-bladder operation for my dad, we had to mortgage the farm. We weren't making enough money to pay off the debt, but somehow we did it. It was a struggle, though. Elijah, my older brother, went to California in 1936 to work on a company farm where he made twenty-five to thirty cents an hour. He agreed to send money back home to help out. Another brother went to a CCC camp and sent us an allotment back home to help out.

In the wintertime, they made cross ties. Using a team of horses, they would haul a load of about eight ties. It would take a week or two to hew out a load of those cross ties and maybe get six to eight dollars a load for two weeks work for two men—just think, ten days labor! They used the cash to buy such things as sugar and flour.

WILLIAM EDWARD DELAMAR

Yes, sir, brother! During the Depression, people got to be very poor, very poor. Now, what a lot of folks don't know but I remember. Sometimes we speak of black people and sometimes you speak of white people down where we were. There were some poor white folks and poor black folks. But down where we were, we would help one another. If I wanted to use your wagon, I could use it. If you wanted to use mine, you could use it. When we'd go to kill hogs, if we needed any help, we would help one another.

S. LORETTA ECHOLS

Yes, but I knew my parents and other family members could handle it. There was a short period of time when we only had chocolate gravy for breakfast. We children loved it—later my mom told us that was all she had to cook for breakfast. I well remember hearing my parents have conversations and be concerned about the food situation. It wasn't scary to me, though, because at that age I just knew my dad would handle it.

But certainly what a bad shape we would have been in if we would not have been able to raise food. But there were a lot of families including some of mine that it was hard to have enough money to buy garden seed, and if it had not been for some of the merchants in town that were very generous and would let us charge it and put it on our credit, people would have been much worse off than they were.

VERNON L. GLENN

Was there a first time when you realized that your family was poor? What caught us is that Daddy would borrow some money to buy seed and fertilizer. But when the Depression hit, you couldn't sell cotton for much. That got us into a bad situation. We had a team of mules—that was the big fad, like a tractor today. Daddy owed some on the mules. But suddenly he had to make a payment on them, which because of the Depression, he couldn't do. So they said we had to bring them back to Smithville where we bought them.

What do you remember about what happened next? One of my memories is that Daddy asked me that day if I wanted go to Smithville—that was about twelve or fifteen miles from where I lived. I was a little boy then, seven or eight. Somebody had to take the mules up there. Of course, I didn't know what was going on. I wanted to go to Smithville. Riding the mules to Smithville was a great thing. But when we got up there, he had to leave the mules and we had to walk back home. That's when it suddenly dawned on me what was happening—we didn't have any mules now and it was about in the middle of crop time. So I think taking the mules back and sorta listening along that line, I figured we were in bad shape.

VIOLET HENSLEY

I never thought about being poor. I didn't know enough about the outside world to know how bad poor I was! If somebody would ask me if I was worried about food to eat, I would say, "I'm just a tomboy girl." I let my dad worry about the food and what we ate and what we wore. I just wanted to learn to do everything he did and I did just about everything. We would patch the horses' harnesses and sharpen our plow points and made knives in the blacksmith shop and I cleaned my own guns and made double-barreled shotgun stocks and I made the .22 barrel gun stocks. You drilled a hole for a bolt to go through to the stock to bolt to the barrel. I've made 30.30 stocks. A double-barreled shotgun is the hardest one to make. After I got older, I wondered where we were gonna get the next sack of flour or the next sugar we wanted. We always had eggs when the chickens were laying.

GEORGIA HEARN HILL

We weren't poor! We had plenty to eat. One time a person asked me, "Did your family ever get plenty enough to eat?" I said, "Yeah!" Oh, my goodness! When Mama got through fixin' up peas and corn bread and butter rolls, English peas and potatoes, collard greens, turnip greens, string beans, all of that, when we left the table our stomachs looked like little tadpoles! [laughing] That's the truth! We had our own buttermilk, our own sweet milk, our own butter. So we weren't poor! A typical breakfast: Oh, my Lord! Ham, biscuits, jelly, molasses, and milk!

MARY FRANCES LOVELL IZARD

We never considered ourselves poor. We knew we were not rich, but the world as we knew it lived much as we lived, and some not as well. Travel was slow, roads were bad, and there were very few cars. Our neighbors around us lived as we did. Some were fortunate enough to find jobs away from home so they could send a little money back to the family.

My sister said she wore her shoes at school but when they started home she took them off and carried them because her feet were more comfortable without them. In our world, if you had something to wear and plenty to eat, what more could you ask? Our young men would soon be called to go to war for their country.

Did your parents ever express concerns about lack of money, etc.? During all the years my dad farmed, I can never remember seeing him overcome with worry even when our crop had just been ruined by a flooding river. Mom and Dad seemed to have an acceptance of life that we do not have today. No matter what problem they had—loss of income, illness, or any other tragedy—they seemed to accept it and go on to try to survive in spite of it.

ROSCOE E. JEFFERSON

We never thought of it as being a hard time. That's all we knowed. You didn't have to have any money! You walked wherever you went. You might come up with a little money along the way to buy a few things, such as salt or soda. A box of soda would cost a nickel. I killed one fox. We skinned it, dried it, and sold the pelt for two dollars.

KENNETH GUY LACY

We did not know we were poor—that was just the way life was. We did realize there were some families who had more than we did. However, there were also families who had less than we did. Somehow, Dad always managed to provide white flour for biscuits, while some families had to eat corn bread for breakfast. We never went to bed hungry as some did.

Did your parents ever express concerns about lack of money, etc.? The only time I remember hearing my parents talk about money was in the spring when Dad would have to go to the bank to borrow money, perhaps fifteen dollars, to make a crop. He always was able to pay it back in the fall.

EVELYN LANGLEY

We didn't know it was hard times. We didn't realize we was having it rough, because that was just the way of life. And that's just the only way we knew. So we just didn't have sense enough to know it could be better, I guess. [laughing]

Us schoolgirls knew we didn't have a lot of things that we coulda used. One year in school during the Depression, the school board members went to Conway and got permission from the Red Cross, I guess it was, and they paid so much to bring these schoolchildren down there and give everyone of 'em a pair of shoes and the boys got a pair of overalls or blue jeans, whichever they was using. The shoes came from an old store that had been down there for no telling how many years. They was old and rotten and didn't last no time. They came all to pieces.

SUZANNE GROSS MARKS

No. Because it was never brought up, to tell you the truth. We just had to get around and do the best we could. The whole town was poor. The grocery store would give us credit; so would the dry-goods store because we had no cash. One day Daddy and I went next door to a cafe to drink a Coke for five cents and Daddy put a nickel in a slot machine and won ten dollars. The nickels were going in every direction and Daddy told me to hold my skirt up near to catch the nickels. What fun to have a lapful of coins!

RUPERT E. "BUSTER" MELTON

I don't know. We never did think about it. [chuckling] Wasn't nothin' good about it. For a lot of years it was hard times. My dad owned a bunch of land. He even loaned money to the poor people.

Was the Depression era a bad time for you? You're dadgum right it was. It was bad for everybody.

Did you know you were in a bad situation? No, I never thought much about it. I just went on and waded through it. I can't figure out how we did survive. But we did.

How did you know the Depression was going on? One of the first things was that you couldn't get a job to work on a farm, and if you did happen to, you would have to work for seventy cents a day. I later went up to Detroit, Michigan, to work in a plant up there. But before that, it was all just working for a farmer.

EVELYN M. COONFIELD METCALF

Almost from the time I was born, I was told, "You can't always have what you want!" We could look at Sears and Montgomery Wards catalogs and dreamed about a lot of those things we wanted, but we had to have money to pay for them. In the late thirties and early forties, Depression was really bad here in Benton County, but we didn't know we were poor because nobody we knew had any money either.

We didn't have credit cards in those days, and most people who did have any money didn't trust the banks. We sometimes had to charge some things at the store in Vaughn, but only the necessities like sugar, coffee, soda, baking power, and sometimes tea.

Making sassafras tea. We found a sassafras tree, dug up the roots, cleaned and chopped them, then boiled them to make sassafras tea. We took corn and wheat to the mills and had our flour and cornmeal processed. We raised everything or did without. When my father paid the bill, I always got a small bag of candy. A real treat!

I got myself into big trouble one time. Although I knew to not charge anything, I just had to have some chocolate drops. My seat was warmed, plus a few other things—I couldn't listen to the radio and had no fun time with my friends. I didn't do that again. We didn't always have everything—maybe what we needed and sometimes not even that. You could buy a lot of groceries in those days with a dollar—if you had the dollar.

Hard times during the Depression. The Depression was an event that probably hit harder here in Arkansas than in some other states. It was devastating for those who lost their jobs, their homes and money. But these were people with great faith, strong minds who did not give in to the problems at hand. There was a shortage of money but not of love, caring, and concern for others. You had to help others in more need than yourself. Neighbors gave of their time, their labors, and their garden, whatever it took to make ends meet and help others.

My father loaned out our horse to the others who had only one horse and needed two. He would take food to Teddy in a bucket so people who used Teddy Bear didn't have to feed him after using Teddy for plowing.

WILLIE MORRIS

Did your parents ever express concerns about lack of money, etc.? They expressed concern but couldn't do anything more during those years. We did the best we could. The plantation we was on was in a way more agreeable with blacks than if we had owned our own land. There weren't too many blacks who owned their own places. Them that did have them lost it because they couldn't pay the taxes or whatever, and were bought out. To tell you the truth, where I was, I learned a whole lot that helped me out up to today, you know—how to appreciate things. So for me, life on a plantation was not all that bad.

LLOYD A. PERRY

You bet we knew we were poor! But we really didn't know what it meant because we had never been rich, and that would have been the opposite to use for comparison. During the "Starvation Years"—1932 to 1936 or 1937—we were always hungry; there was never a time when I *wasn't* hungry.

Sometimes all we had to eat was corn bread and buttermilk for several days, or maybe just corn bread for breakfast. We thought it was special and we were blessed to have biscuits for Sunday morning breakfast.

Wheat shorts, used for hog feed, was about fifty cents a sack (cheaper than bread flour) so we used it to make biscuits. We ate some of our eggs, but saved most of them to trade for soda, salt, and other things we needed but couldn't grow. Eggs brought about five cents a dozen, and a fat hen would bring six or seven cents a pound. We swapped them at the store for necessities. We had to keep our buying limited to our needs, not our wants. People today squander on expensive junk.

We lived by "makin' do" with whatever we could grow and whatever we could make of it. We shelled corn and took it to a mill where we gave a third of it for the milling. Sorghum was important for sweetening, so we always tried to make molasses, and we also gave a third of it to the owner of the mill for his tools and help. If we had anything to spare or that we could do without, if possible, we traded it for something else we needed.

We had the right attitude. I remember a local storekeeper saying, "We're even with the rich. They have ice in summer; we have ice in winter—so we're even up." My mother was able to find a season of work once by moving into Batesville and keeping house for Dr. Gray. She was paid nine dollars a month plus room and keep, really good pay. That helped us out. Our grandparents kept my little brother and me while she was away that winter.

People cared about others and were trustworthy. We have plenty now, but we've lost so much of that. Some have become haughty and overbearing. And many simply don't believe us about the bad times. I tell you, I lived them, and it's nearly impossible to tell anyone how bad those years really were.

UNION H. STOUDAMIRE

Yes, I knowed it! One day I was at the woodpile picking up chips and my cousin came along and give me a dime and I thought I had a big piece of money. I went running inside the house. We didn't have any money, but we raised our own food. We didn't know what poor was. We thought everybody was just like us.

EVA SMOKE WELLS

We knew that some people had things that we did not have but it didn't bother us. We always had fruit and a lot of people didn't, so it all evened out. That's what you call "making do with what you have."

Did your parents ever express concern about lack of money? I don't remember them ever talking about it directly, but I do remember how my dad would sit with his head in his hands trying to figure it out and say, "How am I going to do this?" After my grandfather died and the Depression was getting better but wasn't entirely over, Daddy would borrow money in the spring. Before Granddaddy died, he would furnish the money for the seed and that kind of stuff. But Daddy would borrow money from a bank in Malvern to make his crop and to be paid back when the crop was sold.

And there was always a wonder, *"Am I going to make it because of the weather?"* We wondered if the weather was going to do what it was supposed to so we could make that crop and repay that loan. It was a time when even a child could sense the anxiety that went on. My dad used to smoke Country Gentleman tobacco. It came in little sacks. When he would empty one, he would give it to Mother, who would take it apart, wash it. Maybe she would dye it or maybe she used it as it was. She made a quilt top out of tobacco sacks. My daughter still has it in Tennessee. She latched on to that, saying, "This is an antique!" I don't know where my mother got the dye or what she used but she would dye part of the sacks green. It looked kind of like a brick wall. The tobacco sacks were about seven inches by three and a half inches.

LUCILLE RIDER WILSON

Yes, we knew we were poor and I will tell you why. The other girls whose parents were on the WPA, when they went down to the swimming hole to go swimming, they had bathing suits. We were in the fields working but they were free to go swimming. Their parents had cash money, we didn't.

Did your parents ever express their concern? My father was open to us. Nothing was kept secret. He talked about money with us, about paying debts. We knew exactly where we stood financially at all times. For example, we were kept out of school. I never got to go to school when school started. School had started and we were picking our cotton, which didn't take very long. And then we went to what was called "the bottoms," where we picked cotton for money. That was used to buy school clothes and shoes and that kind of thing. School had started and we were still picking cotton.

FLOYE WINGFIELD

A dozen eggs for a two-cent postage stamp. There was a lot of things you just had to have money for. If you didn't have it, you would just have to do without. That showed up a lot in clothing the children wore. One time, Mama didn't have enough money to mail a letter that required a stamp. She told me to take a dozen eggs to the store and and ask if they would trade them for a stamp. They did and gave me a little piece of candy, which made it worth the trip. I made sure I ate it all up before I got home, too.

JACK WOOD

Yes, we knew we were poor but it didn't bother me. We had enough to eat and Mother made our clothes, most of them. My parents were really upbeat. They just worked on that farm and didn't worry about other things. I'd say the toughest times for us were between 1928 and 1936. People didn't have jobs, people were starving to death. They were lining up by the bakery to get bread that had gone stale. There was just nothing to live on, nothing at all. But we were fortunate because we had this one-hundred-sixty-acre farm down the road.

Chapter 14

Transportation—How We Got Around

DOROTHY COX COSTON ASHLEY

Everyone we knew walked! We walked everywhere we went until about 1935 or 1936 unless we went with somebody else because we didn't have any kind of transportation. This wasn't unusual. It was just the common thing for people to walk from one house to another.

Saturday morning trips to Camden. My grandpa had a Model-T car, one of the very few cars in the community when I was really young. If a trip to town was necessary, anyone in the whole community for miles around was always welcome to ride into town with him any Saturday morning. It was common knowledge that Grandpa was going to leave his house at nine o'clock each and every Saturday morning going into Camden. So anyone wanting to go and showed up at his house were welcome to ride with them. But he didn't wait on anybody. If you told him six times that you wanted to go to town with him and you weren't there at nine, you missed the boat! Daddy bought a Model-A Ford car in 1935 or 1936.

LOUISE J. BRANT

Horses and wagon. Probably took one and a half hours to get to Mulberry, which was in Crawford County. We went in once a week in the wagon to get groceries. Before we left town, we would go into the grocery store and get a dime's worth of cheese and crackers and eat 'em on the way home. Cheese never tasted as so good as it did then. It was a little town. While my parents were shopping, back in those days we could do whatever we wanted to. My favorite uncle and his wife lived just a block out of town so we visited them, too.

GENEVA LOUISE BECKETT COTTON COLLINS

Our transportation consisted of wagons pulled by our horses, a T-Model and a very old buggy. We lived about nine miles from Paris, Arkansas, on top of Calico Mountain. It was a long trip, but we looked forward each weekend to going to town and seeing people. My mom made money from the sale of eggs and butter.

When I was in town, my mother would take me to Sterling's dime store and give me a penny for candy. The clerk would get a bag and fill it with assorted candies. While we were in town, my mom would walk down the street visiting with neighbors. And when she saw a black lady standing with a baby in her arms, she would walk up and ask if she might hold the baby for a while, which the ladies appreciated because their arms were tired.

In 1940, the blacks in Paris, Arkansas, were discouraged from entering stores and the women had to wait for their husbands to buy the supplies. This was very unusual because at the time, white ladies weren't supposed to talk to blacks, let alone to offer to hold their babies. As I always loved babies, I was always eager for this to happen. This influenced how I felt about blacks and I grew up without any prejudices, because of my mother.

WILLIAM EDWARD DELAMAR

Walk or ride a horse! A lot of people would walk twelve to fourteen miles from Dalark to Arkadelphia. You'd get up early in the morning, you'd start off with good daylight, walk over there and walk back late in the afternoon on the same day. We had a wagon so that's the way we got to town. We would bring in the corn to the gristmills, where they would grind it for meal.

As an African American, were you treated well when you got to town on the wagon trips? Yes. I have to say we were. That's one thing I'm proud of. I've never been involved in trouble. I never have been treated too mean. I always knew my stall and that's the one I stayed in. I knew my place.

What other memories do you have of coming to Arkadelphia? When I was just a teenager, I borrowed ten dollars. I went to Arkadelphia and bought me a pair of trousers, shoes, necktie and shirt, and a little ol' cap. And I had some money left. That's how cheap things was. I bought it from a man named Albert Langley. I did pay back that ten dollars.

I never borrowed [more than] fifty dollars in my life. I bought an automobile and made monthly payments. I bought furnishings, and made monthly payments. But just to ask you to lend me twenty-five dollars or thirty dollars, something like that. I wouldn't do it. Working on a farm [as a sharecropper] for someone else, I would get sixty cents a day—sixty cents a day! Most of the time we worked from sunup to sundown. Back in those days, you never paid anyone by the hour. They'd hire you by the day.

ZELMA LOUISE HAMILTON

At one time my dad had three Model-T Fords. Two of the cars were parked under the shed built on the side of the barn. One was near the house. The barn loft was filled with hay, the lower part with corn. The other side of the barn, a shed was used to feed the horses and mules. One night my dad and the boys heard a noise. By the time they got up, the barn was engulfed with flames. We knew someone had set it on fire, but we couldn't prove it. Two of the cars were destroyed. My dad soon sold the other one. After that, it was just wagon and mules.

How long did it take to get to the nearest town? About and hour and a half in the wagon. Moro was five miles away. At times, my dad walked to Moro carrying a twenty-five-pound bag of corn across his shoulder to get it ground into cornmeal.

GEORGIA HEARN HILL

We traveled by wagon. We would go down to my grandmother's in Gum Springs (about five miles south of Arkadelphia) in a wagon pulled either by Nell the horse, who would let us ride her, or old Jack the mule, who was mean. Mama would put us in the back of the wagon with all these quilts and pillows. We'd sleep. It was a good time.

How long did it take to get to the nearest town? Richmond Hill, where we lived, was about fifteen miles from Gum Springs so it took a couple of hours in a wagon. We would start in the morning and get down there before dinner (which is called "lunch" today). After we got there, we would play with my cousins Ruthie, Charlie and Janice Helm. When we were there in the wintertime, we would sit around the fireplace.

Baked cornmeal. Mama Georgie would mix up a batch of cornmeal, rub out a hole in the ashes, pour this cornmeal down into that hole, cover the cornmeal with hot ashes, and let it bake. When she would pick it up, it was brown on the bottom and brown on top. She would break off pieces and let us eat them as we sat around the fire.

KENNETH GUY LACY

The only transportation we had was by horseback, wagon, or by foot. It would usually take four to five hours to go the ten miles to Heber Springs, the nearest town, because of having to let the horses or mules rest from pulling the heavy load. It would usually take all day to go to town, do the necessary business and get back home. During the farming season, Dad usually didn't go to town until the "crop was laid by." After that, he would work in timber, usually getting a load of posts, hammer handles, etc., loaded on the wagon, ready to sell. He would take them to town the next day.

What would you do while you were in town? I remember going to town only once—the year my dad became ill. During that period, his younger brother kinda took over. He had two sons who were about the same ages as my brother and me. It was in the fall. We had a load of cotton to take to town to be ginned. My brother, sisters, and I had helped pick the cotton. He told my brother and me that we had worked hard and he was going to let us go to town with him and his two sons. I remember what a good ride that was burrowed down in the cotton.

I also remember that he bought each of us four boys a maroon corduroy jacket, all alike, that cost two dollars each. That was a lot of money! I don't remember how we passed the time while the cotton was being ginned. I do remember they would bring the ginned cottonseeds home to feed the cow during the winter.

WILLIS MAGBY

Most of the time it was a wagon and team of horses or mules. The little town of Beaton was close by, but you can't find it now. It's in the woods. The town of any size that was nearest was Amity, but we didn't use it. When we did go into town to get groceries, it would be Cedar, Beaton, or Lambert. And there was a dry-goods store in connection with the grocery store in Beaton. It was only a mile or two from the farm, depending on where we were living at the time. Not over two or three miles at the most.

Where did the money come from that you used to buy groceries, etc.? I could make seventy-five cents a day hoeing corn. That was standard. From the cotton we sold or we might work out for somebody else. We'd pick cotton or hoe cotton. I did both. Back during the Depression, I never was a good cotton picker, so I couldn't make much doing that. Some people could make a dollar fifty or two dollars a day pickin' cotton, which is the worst work in the world because it hurts your back! Did you ever pick any?
DOWNS: "No." MAGBY: "I didn't think so." [laughing] No, you bend over to pick cotton and it's a backbreaking job! But there is some people who can do it and it don't bother them.

WILLIE MORRIS

Mule and wagon. It took about forty minutes to go the four miles to town. It was kinda fun. We had a commissary store there on the plantation. But we would go into Dermott just about every weekend.

What did you kids do while you were in town? In the summertime, we would go in there to the Double Duck Ice Cream stand. After eating ice cream, we would go to see western movies, which I think cost just ten cents. Then we would go to one of the Three Chinamans stores in Dermott and get a dime's worth of cheese or bologna, and they would give us the crackers to eat with it.

Wooden wheels on T-Model cars. My first stepdaddy had two old T-Model cars in about 1928, something like that. And you know them things had wooden wheels. When we would get to runnin' around on weekends—we lived right on the Bartholomew Bayou—we'd push them wooden wagon wheels or the car wheels out into the bayou to keep 'em wet where it wouldn't get dry and pop off the cars. When we got ready to get them out, we'd have to get the mules.

LLOYD A. PERRY

I saw my first T-Model Ford in 1925 when I was five years old. The doctor was coming to see my sick mother. It was dusky dark and I saw two big lights coming up the road. There were no mules pulling that thing and I couldn't imagine what it was. It wasn't unusual for those vehicles to scare the daylights out of a team of mules pulling a wagon on the road. They'd take to the woods with you!

EVA SMOKE WELLS

If you had to have transportation, you had to have a wagon. When I was just a few months old, Dad—Rhuben P. Smoke—had to have an emergency appendectomy. Grandpa lived on Smoke Ridge Drive. They had what they called a "hack." It wasn't a buggy, but a light wagon with larger wheels than most wagons have. They took him sixteen miles to the Levy Hospital in Hot Springs in that vehicle with his appendix about to burst. If we needed to go to Hot Springs, we went on the old highway, about eighteen to twenty miles. I never went in the wagon but my uncle had a Model-T Ford and if you didn't have a flat tire or two, then you might get there in one and a half to two hours.

Chapter 15

School Memories

DOROTHY COX COSTON ASHLEY

I started to school in a two-room schoolhouse. One room housed the first four grades and the other room housed grades five through eight. There was a row of seats for each grade. The one I was in had four rows, one for each grade one through four. While I was in the second grade, we moved to Bearden. My older sister and I had to walk about a mile and a half each morning, rain or shine, to catch the school bus to ride to the Locust Bayou School, which had twelve grades.

The school bus had no heater. I don't think any of the buses had heaters back in those days. My best friend and I took turns on real cold days sitting on each other's feet to keep them warm. But for the most part, the school bus ride to and from school was a happy time to socialize with friends and classmates. Each schoolroom had a large wood-burning heater.

There were two outdoor privies located in opposite sides of the portion of the playground that was farthest from the building. The only light we had during the day in classroom time was what was provided by nature.

BESSIE BLACKNALL

I got through the fifth grade. [When asked why she didn't go further in the one-room Mount Morriah schoolhouse, she replied, "It didn't go any further!" (general laughter)]

With five grades in one room, how were the classes taught? In each class, the students would be put in rows from front to back. One row would be the first grade, the next row would be the second grade, the next third, the next fourth, and the next fifth. The teacher would teach one grade at a time while the other students remained in the room. And believe it or not, you can learn something if you were through with your work or you weren't reading or writing. You could learn something from that other class because if you were first grade, you heard the other grades getting their lessons. And if everybody used the same blackboard, you could see them doing their lessons.

Why wouldn't the same system work today? [Rosie, Mrs. Blacknall's daughter, answered: "Because parents don't whip their children. Parents don't have time to fool with their children. I live here on Caddo Street. I can go up the street at any time of the night and I see somebody's junior high school student. While they are out there in the streets at night, where are the parents? Another thing, during my time and my mama's time, if I got into trouble at school, that teacher was going to whup me! And I'd know that when I got home I was going to get another whipping. They call that abuse now but children were better then than they are now."]

And back then, if they went to church, there would be a lady there to teach them. There would be no backtalk, no nothing because that Sunday School teacher would give you a whipping, too. And you would get another one when you got home. You just didn't hear a lot of stuff like you do now. Another thing is that they didn't know about as much as kids do now about drugs and whatever. If you got into trouble, you belonged to everybody. When Hillary Clinton says, "It takes a village to raise a child," I don't know if she knew what she was talking about but that's the way it was then.

FRED NORMAN BLANKENSHIP

I went through the eighth grade twice, not because I was failing but back then, we had what they called "split terms." It was a one-room schoolhouse. We'd have five months in the wintertime and two in the summer. After finishing the eighth grade for the second time, we had a teacher come up and taught a term of school and told my daddy that I was ready to go to high school and that he ought to come down to Melbourne and get me a place to stay. That would have been in 1935. So I completed the twelfth grade here at Melbourne in 1939.

In the grade school, the teacher taught all eight grades, so you know we didn't have much time for every class. We had the old reading, writing, and arithmetic. While one grade was being taught, the other seven grades would be studying their lessons for the next class. The teacher would start each morning with the smallest class and then go up to math, and then science and to geography and spelling.

R. L. "BILL" CARTER

Passed the third grade three times. There for a while at Cypress Creek they went through the ninth grade, but later on, they just went through the sixth grade. I went through the sixth grade two years out there and passed both times. And then they consolidated with Brinkley. And Brinkley thought them little dumb country kids had to go back through the same grade again, so I went through the sixth grade three times and passed every time. [in response to my laughter] That's true!

Back in them days you had two months of summer school so the kids could stay out and gather the crops. We had to walk to school and cross old Wolf Slough. There would be water on each of the narrow dirt roads on the levee that had been built up across there. Them big old water moccasins would just roll off into that water when we would go by there.

Quit school to help family. I went through the seventh grade and passed. I went two weeks in the eighth grade. I was so much bigger than the other kids. My family needed my help in trying to make a livin' so I went two weeks and quit and started sawing logs with a cross-cut saw in a steam-powered sawmill. Then I was called into the navy in 1943 while I was sawing them logs.

School play. I remember one time we had three rooms in that old schoolhouse. There was also a stage for school plays. Me and two more boys, we was in the first grade or "primer," they called it back then. Anyway, they had a tub sittin' on the stage. Each one of us had a fishing cane. One of us would say, "Oh, I caught a fish that weighed ten pounds!" Evan said, "How do you know it weighed ten pounds?" And I said, "Because it had scales on its back." The audience just clapped and laughed.

DOYLE L. COLLINS

Possom Trot was a rural, one-room school for all eight grades. There was a stage up front with a blackboard behind the teacher. In front of the stage was a long "recitation" bench. The teacher might call out "First-grade reading" and all of us first-graders would go sit on that bench and have our reading class. I attended Possom Trot School until the fifth grade when it was consolidated with Ash Flat.

We started to school in mid-July after the crops were laid by. We would have school for about eight weeks, dismissed to gather crops, then returned for four or five more months of schooling. We'd have about seven months for the entire school year.

Sometimes we would start our morning off with some Christian hymns—even though this was a public school—led by the teacher. There was a wood-burning, potbellied stove in center of the classroom. We brought lunch from home, which might consist of a fried egg or a fried potato in a biscuit. Mom would sometimes also pack a fried apple or fried peach pie. Mom would sometimes ask one of my older brothers to bring us a hot meal to school.

WILLIAM EDWARD DELAMAR

I only went to the third grade. Oh, it was just one room and it was small. It might have been about twice as large as this room, probably twelve feet by thirty feet. The boys sat on one side and the girls sat on the other. They all sit together now. Back in those segregated days, all the students were black in my school. We had to walk to walk to school on the highway. When white children were riding school buses to their schools, we had to walk. I didn't know what a raincoat was. If it was raining, I just kept on walking. If the water was across the road, we would wade it. And if a bus come along, I had to step aside to keep the water from being splashed on me.

I can remember when we went outside to play baseball. Back then, we made our own balls. We would unravel a worn-out sock, get us a rock and cover it over with a rag. Then you take the string you're pulling out of the old sock and start wrapping it around that rock. When you get through with it, you tie it down. Then we'd get a baseball bat made out of anything—we'd pick up something like a one-by-four plank, cut it out, and make a handle on the end of it. Kinda like a paddle on a boat, you see? And then we would start playing.

Sometimes we'd go out on the edge of the woods and pick blackberries, huckleberries, dewberries, muscadines, chinquapins . . . all of them things. They were like acorns. But you don't find them anymore.

GENEVA KING EMERSON

For the first five years and the summer of the sixth grade, I walked less than a half mile to a one-room country school. The building had three windows along each side, a stage for performing—both the plays I loved and the demonstrations at the blackboard of knowledge gained. The "recitation" bench sat up front by the teacher's desk. When she called us to it, we'd better be able to recite what we had learned from our assignments—staying in at recess and after school or notes home weren't pleasant. In those days, a child who would not study and behave in school was apt to find his life's calling in fieldwork.

A line of homemade wooden seats with desks on the back ran along each side of the woodstove in the middle of the room. At the back, there was a bench where the water bucket with the common dipper sat. Honor students were allowed to go by twos to fetch water from the spring 'way down the rocky hill in the woods. Our "library" was one small cabinet with discarded textbooks and a set of encyclopedias proudly displayed. Probably the latter were obtained with proceeds from a pie supper.

BEULAH LEE MCLEOD EVANS

We attended school two months in summer and six months in winter amd worked in the fields during planting and harvest times. We walked two miles to school, crawled over fences, and crossed creeks. A lot of times, we'd slide off and our feet would get cold. I remember one time the ice was on and I slid off a rock. My feet were cold and [my teacher] Miss Moore peeled my sock and shoe off and got a chair and set me up close to the stove and pulled my sock and shoe off to dry out.

MARY EATHEL FREEMAN

I walked three miles to a bus route and rode a cow truck with a canvas cover in bad weather. We enjoyed about six months of classes each year. School was a time when we saw the neighbor children every day. It was a time to relax from field crop work—a vacation from farmwork.

During the second semester [of my ninth-grade year] we had literature. I was so happy—all those stories. We read *The House of Seven Gables* by Hawthorne, *Tom Sawyer* by Mark Twain and *Ichabod Crane* by Washington Irving. My favorite poems were "Thanatopsis," "Lady of the Lake," "To a Waterfowl," and "The Psalm of Life." [Miss Nation] my teacher saw that I was fascinated. She kept recommending books. I haunted the library and simply could not get enough reading. I may have used it partly as an escape.

[Author's note: Excerpt from Eathel Freeman, *Journey of a Depression Kid,* 3rd printing (self-published, 2005), 21. Used with permission.]

VERNON L. GLENN

I think my dad went through the third grade and my mother—a little bit better than that. But, they insisted they go to school, rain or shine. My dad, who went through the third grade, had a formula for whether we wanted to go to school. If we didn't want to go, then we could work on the farm. We figured that pretty fast. I was younger than my brothers. We would all go together and then a whole bunch of kids would go join us on their way to school. When it was over, we would all walk home together. I learned a lot of things walking with those older boys.

We started in a schoolhouse that had three rooms that were for everyone from the first grade all the way up. The WPA came in and was gonna build a school so they tore that old one down and we had school in the Methodist Church. We didn't have a well so we would go to a fella's well and carry water so we would have something to drink at school. But then they were trying to teach us hygiene like how to keep from getting a disease by drinking out of the same dipper.

We had this little tale about this one little boy who told his mother, "We're having hygiene." She wrote to the teacher and said, "Please send me a book on hygiene, I think my kid has it." That was a great joke that went around for a long time.

No restrooms in the school. The way we handled it was that there was a pretty deep ditch that water had washed out over a period of time. The boys went down there on one end and the girls went up here on the other. But you know we never thought much about it; that's just the way it was.

VIOLET HENSLEY

We had a one-room schoolhouse, which was near the church house. We had to walk barefoot until our toes got so cold. We didn't have shoes until we sold the cotton. If it was raining or if it was too cold, Dad kept us out, so I went to school up to the seventh grade. I liked reading, writing, and spelling. We had spelling bees. I was usually the last one to sit down on spelling.

We didn't go to school all that much because we had to pick cotton and gather corn in the fall. And then in the middle of winter if it was raining we couldn't go because we had to walk two and a half miles. We had to cross the creek on a foot log. If the creek would wash the foot log away, we'd have to go cut a new tree and let it fall across the creek. When the water's up and muddy, you'd better watch out or your head would be going down the creek with that water.

GEORGIA HEARN HILL

I went through the eleventh grade. We went to Richmond Hill, an all-black one-room schoolhouse, until the eighth grade. On a Friday at Richmond Hill, we would have a spelling match and I was always the one who was head of it. Discipline was never a problem. We didn't do anything like children today. No! Chewing gum? Walking around and talking? No, sir! We were very, very quiet.

Today, children have lost respect. It started at home, letting children do what they wanted to do. That's why it has gotten out of hand. When prayer was taken out of the public school, the result was chaos. When we started going to school in Arkadelphia, I had to walk about two miles from my house to the highway to catch the ride with my relative, Miss Margie. We didn't have any discrimination between white and black kids. The white children had a bus, but at that time, we didn't.

MARY FRANCES LOVELL IZARD

I covered the first through the eighth grade in two rooms. The walk through the woods from our house to the school was two to three miles. All of the students, boys and girls, were barefoot except one. There was one creek that did not have a bridge. Cars or animals forded the creek but pedestrians crossed on a footlog. The footlog consisted of a large log that had been flattened a little with an axe then laid across the creek. Walking that footlog was my greatest nightmare.

I felt inferior to students who lived in town and didn't have to ride the school bus. They had store-bought clothes and could go home at lunch while I ate my lunch of a hard-cooked fried egg on one of Mama's biscuits. This lunch was rolled up in newspaper because we had no paper bags and could not afford a lunch box. It didn't take me long to realize that I might not have a lot of the things they could afford but I could do one thing—make good grades!

ROSCOE E. JEFFERSON

I went up to the third grade. I quit because we had to work. Back in them days, school was just three months at a time. Three months in the spring and three months in the winter. In crop times, we was at home working. We had to hoe a lot of cotton, and when fall come, we had to pick it.

KENNETH GUY LACY

I walked to a school, which was about one-half mile from where we lived, through the eighth grade. That was a one-room school, with one teacher for all eight grades. To start high school, which was twelve and a half miles away, on a dirt road, I rode in the back of a truck that had a wooden box built on, with a door on one side and a small window in the back. There was a wooden bench on each side to sit on.

The accident that ended my formal education: Water was obtained from a well, on the school grounds using a metal hand pump. A cousin and I were "horse'n" around, when he pushed my head down hitting my mouth on the pump handle, breaking a front tooth. It became abcessed and had to be pulled.

Since it was in the front, the dentist made an actual tooth and put a gold crown on it. It cost my dad thirteen dollars. Later the tooth was taken out. I still have it and the gold. After the accident, Mom decided it was time for me to go to work on the family farm.

DULAS L. MASSEY

Our school had no electricity, no running water. We all drank out of the same bucket. Some of us had our own folding cups. Restrooms were in special areas down in the brush, one for girls and one for boys. We didn't have any restrooms or electric lights or running water, either. Kids would get a two-gallon bucket full of water and go over to a neighbor's house and put it on a stool in the schoolhouse with a dipper. If you wanted a drink, that's the way you got it.

When you had to go to the restroom or toilet, how could you wash your hands? There was no ways to wash your hands. And you didn't have no toilet paper, either. You'd just pull some leaves off the trees, strip off the lower leaves. We just didn't think about it.

GEORGE MERTENS

My first five grades of school were in a one-room schoolhouse—with one teacher—in the Idlewild Public School (Prairie County). The building still stands today. Many of the kids were really poor with no shoes and ragged clothes. Some went barefooted all winter long. They developed such tough feet that when the road ditches would freeze over and we would slide on the ice, they would slide right along with us. We would bring our lunch. Some would have very little to eat, maybe a biscuit and some cornmeal mush. If one of us had an orange, they would ask for the peeling.

EVELYN M. COONFIELD METCALF

I got through the eleventh grade—one at Eagle Corner, six at Vaughn, four at Ulysses in Benton County. Eagle Corner School was built in 1893 on one acre of land that cost fifteen dollars. We had a small library at school. We would take the book home to read. *Bobby's Twin* and *Little House on the Prairie* were great.

We dressed so much differently then than we do today. Even us girls wore long-handles and long cotton sacks over our knees. We wore a lot of clothes to keep us warm. When we got to school, it wasn't always real warm. And it took a while for the old woodstove to start putting out enough heat. We would carry our lunch in a syrup can or bucket. It might contain biscuits and eggs or maybe some bacon. Maybe cookies if there was some. My mother always tried to have cake, pie, or cookies for my lunch.

Sometimes our teacher would tell us she would be bringing a soup bone to make soup. We could bring some canned vegetables or onions, maybe tomatoes, and put everything together. We brought a cup or fruit jar to put our soup in and a spoon. Someone usually brought one or two loaves of sliced bread. Everyone got to eat. If students didn't have anything to bring, that was okay. We all understood. We studied hard, played hard, and enjoyed each other.

WILLIE MORRIS

Back then, we attended a three-room school for blacks—two classrooms and one room for the principal—for grades one through eight. After that, we went into Dermott, where I later finished high school. My stepfather didn't want to take us out of school to work on the farm because he said he would see to it that we each got through the eighth grade.

We walked to and from school. About a quarter of a mile. After the school was relocated, maybe we walked a half mile. We had to come from one side of the bayou to cross the Bayou Bartholomew Bridge to the school. so that would take us a little longer. We walked four miles from Baxter to Dermott, where I went to high school. To cross the bayou and get to school, we used a raft made out of logs. We'd take a wood paddle or something like that and paddle across to where we could make our shortcut to school. That was kind of hard on a cold or rainy day but we got on across. Later on we had a school bus.

LLOYD A. PERRY

Eighth grade was as high as we could go at Curry Springs (Sharp County). That cut off education for most kids because they'd have to rent in town to go to high school. But a student who could pass that eighth-grade test had an education that some college students couldn't match today. We walked to school, some two or three miles. If the weather was too bad, we didn't have school, partly because of that. The only heat was a woodstove that set in the middle of the room.

As for games, we played dodge ball, tag, and "ante over." This game was played by tossing a ball over a small building with a V-shaped roof clear over the roof and not hit the shingles. We'd call out "Ante, Ante, Over the Shanty" so kids on the other side of the building knew the ball was coming. But we would throw it at an angle to surprise them and make it harder to catch. If someone from over there didn't call back pretty quick, we knew the ball had been caught and we prepared to try to run to the other side without getting hit with the ball as they came after us.

WILLIAM PIERCY

The old Vaughn schoolhouse is still standing. It was a two-story concrete-block building. Four rooms downstairs, grades one through eight; four rooms upstairs, grades nine through twelve upstairs. It was heated by an old coal furnace. No air conditioning.

School had three buses but no heaters in them. Students had to live more than two miles from the school in order to ride the bus. Bill remembers there was one family that had so little, that all their kids had to eat for lunch was fried potatoes from the night before. Because there was no heat on the bus during the winter, the potatoes would freeze before the kids got to school. Bill has never forgotten those kids sucking on their frozen fried potatoes.

[Author's note: From notes made by his wife during the interview.]

ALMA POUNDS

We had some one-room schoolhouses but mostly it was two rooms. I started in a two-room school at Lunceford (Arkansas County). We had double desks—two kids to a desk. Everybody took care of everybody else in school. If I knew a lesson and could help someone else I had to help them. One summer I went to a place called "Good Luck," which was on the corner of one of granddaddy's farms. They had all eight grades in one room. The teacher would have the eighth graders teach the little ones. Not just constantly but different things.

How could eight grades exist in one room? There weren't that many people, maybe twenty to thirty. But we would go into a corner and have class. Or we could sit and listen to the higher grades if we weren't working.

JOANNE RIFE

How the way we had lunch affected two groups of people at the school in Vaughn: At home we had dinner at noon and then supper at night. I had a lunchbox with a thermos bottle in the top. I had hot soup or cold milk. I had sandwiches and fruit and cookies, a variety. Leon McGaugh had fried chicken in his lunch and thus earned his lifelong nickname "Dead Chicken." But one family sent a quart jar of canned peaches with the girl in the fifth grade or so; and the two little boys, her younger brothers, brought a "poke" with cold biscuits from breakfast, and that was their lunch.

UNION H. STOUDAMIRE

I only got to the seventh grade. I walked to school, which was six miles away from home. It was a one-room school with a one-pupil bench. I did not have books. My uncle had a bad ulcer sore on his leg and it took what little money he could make to buy his medicine for it.

I had to take the role of a housekeeper for my uncle. I went to school every day except some days I would have to stay home to help my uncle to get wood. I would be glad when I got to go to school so I would have a chance to get to play with my friends. Our school was a one-room wood building with just one teacher. He was a man who had four sons. They all went to school there. His name was Professor William Henry Barnes. He was called W. H. Barnes. He was a good teacher. We had school at Barnes School in the winter.

Mount Olive school, also a one-room wood building, had one teacher, a female. Very good. We had to take our dinner every day. Some time we would swap something we had with each other. We had a water keg with a tin cup to drink out of. Everyone drank out of the same cup. Our seats was cute little fold-up seats. We had a large blackboard in our school. Sometime the teacher would make us work our lesson on the blackboard.

JAMES A. THOMPSON

I guess I got up to the fifth or sixth grade in a one-room schoolhouse in the little town of Lurton (Newton County), Arkansas. We had one teacher and he taught all the different grades, four or five. We all went in and sat down. One class would sit in one portion of the room. Another class would sit in another. While one class would be in session, we'd just sit there and wonder what we were going to do when we got home.

What kind of sack lunch did your mother fix for you? Probably a biscuit—we didn't have bread like we have now. And maybe—and that's a big maybe—a piece of meat like sausage or a piece of ham. We didn't have much of that. And she would put a bunch of vegetables. Some of it was raw, some of it she would put into a pot or pan. We just didn't have much to eat for lunch. However, everybody was in the same shoes. I was no better off than my neighbor next door.

How would you keep the schoolhouse warm on cold winter days? We had a big iron stove in the schoolhouse. The teacher would generally get there before we would in the mornings. She would build a fire and get it to going. When we would come in, we'd throw another log or two on the fire and you know? That old stove would get nearly red hot sometimes. The little girls didn't have many clothes on because they didn't have many clothes back then. They would back up to that stove and I'm telling you what, their little legs got red. It heated up the schoolhouse. We were all pretty tough back then. It was just as good as anyone else had.

CHARLES WHITFIELD

I think I was in the fifth grade when I quit. I wasn't too interested in school. I was more interested in making something to live on. I didn't have to quit. I quit to help myself. I wanted to make a dollar. I didn't have no money to spend. I got a chance to work, so I decided to work instead of going to school. That was another mistake. When you quit school, you made a mistake.

ALBERT M. WILLIAMS

Did you ever go barefooted? Yes, and I got a lot of scars on my feet to prove it. When I started going to school at Monticello, about four miles or better. I walked barefooted in the snow many times. *Weren't your feet getting really cold?* No, they were freezing! [laughing] I got frostbitten feet and toes many times because of that cold weather! They stayed sore a long time.

Why didn't you take the school bus? School bus? Who had school buses? I never heard of that back then.

LUCILLE RIDER WILSON

During the Depression, we were a poor family, in a poor school district, in a poor state! Because we weren't able to pay the tuition the first six weeks, we entered late when it became free. That means we attended about six and a half months a year. I couldn't be present when school started because I was picking cotton for money in "the bottoms." The money was used to buy school clothes and shoes and that kind of thing. School had started and we were still picking cotton.

The schoolhouse. The first floor was the first six grades. I remember how much I loved to read. I didn't have books, I didn't have crayons, I didn't have scissors, those kinds of things that kids have now. But when I got to school I found the library, which consisted of two or three book shelves. I remember one year when I found twenty or so books on people being "saved." I knew that was not appropriate to be in a classroom but I read every one of them. Reading was my escape and my contact with people, places, and how other people lived. I read every book.

JACK WOOD

The Clark County Poor Farm. As we walked each day we passed by the Poor Farm run by Mr. Doc Greeson at Gum Springs and all the poor people came to the fence and wanted to talk to us. We kept on walking slowly because they were mentally unstable and they wanted us to help them. Made us feel sad each time they came out to the fence. I understand that they were worth a dollar fifty-seven per day, each paid to Mr. Greeson daily for their upkeep. The fee was paid by the government.

We got our first car, I believe it was a two-seater 1932 Ford. It had curtains on the side that you could unsnap to get them down. So I drove the kids to school in Arkadelphia where we all graduated.

[Author's note: Established in 1887, the Clark County Poor Farm was one of five such facilities in Arkansas (others were in Benton, Carroll, Clay, and Cleburne counties). Located about four miles south of Arkadelphia on Highway 67, its purpose was to help those who could not help themselves because of age or disability.]

Chapter 16

How Did We Entertain Ourselves?

DOROTHY COX COSTON ASHLEY

Catching whitefaced carpenter bees and horseflies. My friends and family cannot believe that I actually did this and tied a string around their necks so I could play with them as they flew around. But I did this often until I got old enough for it to be boring to me.

We made paper dolls from pages of the Sears & Roebuck catalog. We even made paper furniture for our paper dolls. For the most part marbles was considered a boys' game. I had a few marbles, but they were unimportant to me. We made up parlor games such as club fist, Button, button, who has got the button? and I spy.

We played with the baby animals on the farm. Our family kept several litters of kittens and we always played with them. In the summer months we would go into the nearby woods and rake out squares to resemble rooms of a house and we would play house. The person being the "baby" was always acting up and crying and the "mother" was always spanking the "baby."

Making frogs smoke cigarettes. We played various pranks on each other. A favorite prank of the older boys was to make a frog smoke a cigarette. What's that? There is something about the texture of a frog's mouth that if you put a cigarette in it, the frog can't spit it out so as he tries to spit it out, he puffs on it. We had to make up what we were going to do.

"Who, what, when and where?" One year we played this game with some friends. A different person whispered into your ear who you were with, what you were doing, when you were doing it, and where you were doing it. Since no one knew what the other people were telling you, we came up with some pretty hilarious stories when we told everyone what had been whispered to us. We never knew what kind of story we were going to have until we had been told.

Most of our games required imagination because we didn't have any equipment. My sisters and I received a Chinese checkers games and some dominoes games one Christmas and we also received a deck of cards one year. We played Spades and Old Maid and what we called Puppy Feet—which were clubs.

KATHRYN BAJOREK

When my little grandson was over here one day, he asked, "Nana, what did you do for fun when you were my age?" "Well, I picked cotton a lot," I replied. "But I remember one Sunday, my daddy took an old stick and he hammered a lid off of a bucket, nailed it to a stick, and told us we could run that down a dirt road and make a track."

"What?" my grandson asked, having no idea what I was talking about.

"Then he took an old KC Baking Powder can and ran a wire through it and filled it with dirt. We pulled the wire and it made like a wagon track. Wasn't that cool?" I asked my grandson.

"What are you talking about?"

"Honey, I didn't have any shoes." My grandkids have had everything like laptop computers of their own since they were in the third grade, and are as smart as they can be. "When I was your age," I said, "I never had any shoes." He looked down at his shoes and eased his little ol' feet out of them and said, "Well, I'm trying to get my feet tough by walking on the gravel on the driveway."

"That's okay," I said, "you don't have to go without shoes."

Still not satisfied, he asked, "What did you do for fun?"

"We split wood," I said. "And we jumped stumps. Whoever jumped the most stumps without falling off were the winners. Stumps were what we would get when we were clearing new ground and you have stumps sticking up that we would try to jump. That's why we called it 'Stump Jumping.'"

"I just can't believe you did that!" he said.

"Honey," I said, "that's all we had. We didn't have a radio and a TV. You know, I think I was born to the wrong family."

"Oh, don't ever say that!" he said. "If you weren't born to our family, you wouldn't be who you are today." I thought, "That's a pretty good answer!"

BESSIE BLACKNALL

We would play hiding-the-bee-box (similar to hide-and-seek), A tisket, a tasket, etc. We'd play marbles, make swings in trees, played baseball. To make balls, we'd get a little bitty rock and a bunch of rags and sew them together. We'd use a stick for a bat. We'd also play London Bridge and Little Sally Saucer. We'd make a big circle of people holding hands. One child in the middle. Then we would start singing:

> Little Sally Saucer setting in the water
> Rise, Sally, rise. Wipe your weeping eyes.

Put your hands on your hip, let your backbone slip.
Shake it to the east, shake it to the west,
Shake it to the one that you love the best.

You'd pick out someone in the circle that you loved the best. And when you shook the saucer at that person, you'd take that person's place who would get inside the circle and do the same thing. The fun of the game was picking out the person you were going to love, etc.

LOUISE J. BRANT

We had to make our own entertainment. The boys fished and hunted. For entertainment, we had a family that lived a mile south and another mile east of us. Our social activity was going to the Price's play party on Saturday night. This is the activity that we looked forward to. We played games, pulled taffy, simple things like that. Had an old Victrola and played records on it. For music, some of the people in the community had Jew's harps, fiddles, along with horseshoes, balls, checkers, cards, dominoes, etc. My brother Jack won all the good marbles in the community. [chuckling]

MARY LEONA CARTILLAR

We played pitch. I was a good pitch player. That's where you play with cards—like poker. But, we played pitch. I was Daddy's partner. Of course, I knew what Daddy had in his hands at all times—we used signs. And we always won anyway. We played with the neighbors, too. We had a good time. We had kerosene lamps but no electricity. Sometimes we played way after midnight.

DOYLE L. COLLINS

Well, we played marbles; Town Ball, which was played with a rubber ball. You had to catch it on the fly or on the first bounce. It was an offshoot of baseball. We'd also play softball, tag, kick the can, red rover, farmer in the dell. Whatever we played couldn't involve too much equipment. Indoor games were checkers and dominoes. We also had community entertainment. This included listening to small string bands with guitars, fiddles, etc. They were composed of local people. We also had Singing Schools, community singing parties and other parties for local youth, and school programs.

GENEVA LOUISE BECKETT COTTON COLLINS

As the youngest, I had time for play and fun. As for games, we never wanted for things to enjoy. Each Christmas, we got a gift and the rest of the time we played with what we had. Dad was very good at carving. He could take an empty spool, after Mom had used up the thread, and carve it. Then, with a burnt match stem and a rubber band he cut from a worn-out inner tube, he would fashion things one could wind up and watch it go. A piece of string threaded through both holes in an old button, threaded on middle fingers, whirled around, then pulled over and over would make a very pleasant sound.

Dad made stilts and a merry-go-round. He could take two long pieces of lumber, attach a block of wood on each long piece, and make stilts to walk on. For the merry-go-round, he would cut down a small tree at the edge of our yard, then with a large round metal spike through a two-by-six in the middle and small wood strips on each end for us to hold on to. Then all we needed was for someone to push us around and around.

Daddy's downhill cable ride. Our house was across the road from a pretty steep little hill. Daddy nailed an old pair of metal uncovered bed springs to a tree down the hill. Up the hill, he attached a pulley from a water well and also on the tree at the bottom. He threaded finger-size metal cables, attached the cable with hand-holds on it. The point of this was to be up hill, grab hold of the pulley handles, kick toward the downhill tree, lift one's feet, and fly down the hill to hit on the bed springs on the tree at the bottom of the hill.

ARTEMEZE EDWARDS

We played with rag balls, pitched horseshoes. We also climbed trees, made tree houses, raced, shot marbles, played hand-clapping games—you would beat out a rhythm by a person standing in front of you clapping her hands to rhymes like:

Mary Mac, dressed in black, she sewed buttonholes down her back,
Asked her mother for fifteen cents to see that elephant jump that fence.
Jumped so high, darkened the sky, Never got back till the fourth of July.

It was fun if you didn't have anything else to do. You could also do this when you were jumping the rope. We also played hopscotch, hide-and-seek. It sure was a lot of fun.

GENEVA KING EMERSON

We played marbles, horseshoes, stilts, roll-the-hoop, tag, etc., plus helping with the new baby. Learning needlework, carrying items to and for the old folks. Part of the price for playing with babies—and the honor prize for "growing up"—was taking care of them while the parents worked and/or visited (often done at the same time). That meant washing up spit-back and changing yucky diapers (no throwaway diapers or wet wipes then, either). If the little one had grown somewhat, he or she might be a mean little thing, but we couldn't retaliate if we were struck or bitten because we were "big" kids and knew better.

Taking care of our elders in the community was a matter of duty and pride, mixed in with love and respect. No request they made was to be ignored if it could be accomplished. Most of them would just try to do without before asking already weary, burdened family for favors. We tried to watch and ask what was needed and desired.

Earning unspoken honor and trust. One time, when I was about eight or nine years old, my dad sent me a couple of miles over the woods trail to the nearest local store to buy a box of soda for my bedfast great-uncle and -aunt who couldn't leave him to do the errand herself. When I returned, Dad sent me a mile over another wooded trail and across fields to take the soda to the sick uncle. I was delighted to be so much needed and appreciated. It had earned me some unspoken honor and trust.

Every scrap of anything was saved to make quilt blocks, etc. We cut off the tails of otherwise worn-out shirts and dresses to piece into quilt blocks as we sat by the fireplace during winter storms or at night. Blocks were sewn together to make a top, feed sacks had been washed and dyed with walnut hulls for a lining, de-seeded and carded cotton saved from the fall picking was layered between top and lining, and those were fastened with thread or tacks to homemade quilting frames in which the material could be kept taut for stitching all materials together. I learned to piece quilt blocks by school age and do the actual quilting before my teen years.

LaVERNE WILLIAMS FEASTER

There were many fun things for youth to do on the farm. Boys played marbles, rode their bikes, climbed trees, shot slingshots at birds, and fished in the ponds and ditches that were close to their homes. The boys would get together and decide what they were going to do and where. They then let parents know where they were going to be. They'd be climbing trees in the woods, visiting in each others' homes or down at the train trestle. Boys would come down and watch the train come through town.

As the only girl, I was taught to play with dolls. I had tea tables and dishes. I never owned a bicycle. The boys rode bikes, shot marbles, ran back and forth to each other's homes. Girls were taught to be "ladies," not people in the outdoors. The girls were at home. We helped clean, cook, wash, and sew.

However, we had fun playing with dolls, collecting trinkets, playing games, dressing up our dolls, and having play tea parties.

ROBERT FLANNIGAN SR. • S. LORETTA ECHOLS (Joint interview)

FLANNIGAN: We had a lot of parties back in those days, you know. I never will forget one day somebody asked me, "You wanna go to a party? We're having a 'tacky party'" and it was about three miles away. I didn't know what a "tacky party" was. They told me and I got all fixed up and we walked three miles over there to that party and three miles back. We found out that a "tacky party" was to dress up tacky!

[Author's note: Haley Smith, the interviewer, said, "We still do that here at Buffalo Island High School when we do our homecomings and stuff. We'll come up with dress-up days and we'll dress tacky."]

We did things together. When I was about fourteen, we went swimming out here at Cockleburr Ditch and over there in the St. Francis River. We went over there and swam, fished and did things like that together. We had entertainment. We always found something to do, like sandlot baseball.

ECHOLS: We got together a lot and I think that was one of the things that helped get people through [the Depression] is they were not isolated usually and the families were not far from one another. We had cousins and neighbors all around. We also had music. We didn't have a radio, but there was always someone in the community that played the guitar or the fiddle, or what they'd call the Jew's harp or harmonicas.

Then we would have a lot of games. We didn't have to have a lot of toys. And we lived in an area that was farmland but there was a lot of woods where you would spend half of our day playing. Now we had rules that we had to go by and my older sister would always be the one who made sure we were all safe. And we'd have Bayou Meto, a waterway that runs through a lot of Arkansas. It had snakes in a lot of places, so we had a lot of strict rules. We'd play in the bayou, collect mussel shells, and that was our entertainment.

BEULAH LEE MCLEOD EVANS

We had a lot of parties where we played games like spin the bottle and ring-around-the-rosy. In spin the bottle—we'd get to go around the house with somebody a lot of times. We thought we was really big. Sometimes the boy would be bigger than we were. We thought we were really big getting to get to go around the house with somebody. We would spend the night with friends that was having the parties, and when they was having one close to our house, they'd spend the night with us. Most of the traveling was by foot.

We hunted rabbits and rode broomsticks. Well, my younger brother Don made me a slingshot and I hunted rabbits. Instead of bicycles to ride, we would take a broom handle, a red Prince Albert tobacco can. We would take them apart, bend them, and lay them on a broomstick or whatever you had, and you'd just go all over the place. You just thought you was riding everywhere. The thing was to keep from letting it go!

MARY EATHEL FREEMAN

During the summer of 1931, my sister and I played with paper dolls in the hall of the old house. We had many doll families cut from Sears & Roebuck and Montgomery Ward mail-order catalogs. We made furniture from match boxes and other cardboard. It kept us entertained for hours. Later, we played wolf over the river. We chose sides. Two rings were drawn and ten sticks placed in each ring. We tried to steal the sticks by crossing the middle line. The one who got all the sticks won. We also played dodge ball, built play houses, gathered wildflowers, chewed sweet gum, ran races. By the seventh grade I could run fast and hit well.

[Author's note: Excerpt from Eathel Freeman, *Journey of a Depression Kid,* 3rd printing (self-published, 2005), 14. Used with permission.]

ZELMA LOUISE HAMILTON

We played on our merry-go-round—which was a board with a railroad spike hammered down into a stump. It wasn't a smooth board, either! Girls couldn't wear blue jeans back then so I had to wear a dress, which meant I picked up a lot of splinters.

Pop the whip: A number of children, usually ten or more, would line up holding hands. The one at the head of the line was always the biggest one. The one on the end was the smallest. All the children would run. The one at the head would turn and pop the other children. They had got up quite a speed. The small one often times got a few skins and bruises.

Wire briar limber lock: In this game the children would sit down and place their hands and fingers on a flat surface. They drew straws to see who would count. The one counting would say, "Wire, Briar, Limber Lock, three geese in a flock. One flew east, one flew west, one flew over the coo-coo's nest. O-U-T spells out goes he!" The finger that was landed on was turned under. The one whose fingers were all counted out, won. Then he got to be the counter. The one who was counting could no longer play until all the children were counted out and given a chance to count.

GEORGIA HEARN HILL

My cousins made homemade wagons. They would cut a pine tree, cut the wheels, and then take a hot iron and bore a hole in there for the axle. We just had a good time. We didn't know nothing about little red wagons! [laughing] I remember this boy who said, "You can't play marbles on Sunday!" We weren't allowed to play jacks, or play cards or dominoes. I don't know how to play any of that now.

Were you allowed to dance? Not when Mama and Daddy were home. And I never did learn how to dance.

MARY FRANCES LOVELL IZARD

All you need for a game of hopscotch is dirt to draw it in and a piece of broken glass to throw. We played lots of hopscotch. I also loved playing ball and jacks and jump rope. Sometimes we would make our own paper dolls and then make clothes for them. The boys played marbles. We never had many store-bought toys but a child has a vivid imagination and a piece of glass or a pretty rock can turn into anything.

In the summer we went to the creek to swim, swing on grapevines. We just wore our dresses. There was no one around to see us. But if anyone saw a snake or thought they saw a snake, we would go screaming to the house. We had some cousins who lived not very far away and they had a wild grapevine swing. You have to realize that our hills were very steep and the valleys were narrow. This grapevine was attached to a big tree at the top of a hill. If you held the grapevine and took a running start you went so high out over the valley that you felt like you were flying. We made our own fun wherever we were.

As a girl, I played house. The house consisted of rocks laid side by side to form rooms. The dishes were often bits of broken glass we had found. It was a time when children used their imagination and the imagination has no limits. Children did not require purchased toys to have a good time. We also liked to listen to our parents and other grown-ups tell stories of what happened as they grew up. My dad could entertain us for hours telling stories.

KENNETH GUY LACY

The only toys we had were what we made from sticks, boards, tin cans, etc. We had a ball made of string that we would throw to each other, maybe bat it, if we could find a piece of an old board. Someimes we would tie a rope to the end of a board to make a type of sled. One of us would sit on the "sled" and another would pull it. Usually the older kids would pull while the younger ones would get to ride. We bent a tobacco can into a "u-shape," nailed it on one end of a stick, and used a lard bucket lid as a wheel, putting the lid inside the tobacco can to roll it.

EVELYN LANGLEY

Community musicals. I was still a young child then. We'd have community musicals at somebody's house every week. The men who could pick the guitar and play the fiddle would meet, put their chairs out in the middle of the floor, then with their guitars and everything, they'd make music, and we'd sing, if we wanted to, or some of 'em would dance.

Another thing we done, on certain nights, we'd have pea-shelling nights and invite all the neighbors and boys and girls. We'd all shell peas for an hour and a half. We'd stop shelling peas till another night and then we'd have this party and the boys would bring their guitars. We'd make music, just dance and show out. We made our own pastime.

WILLIS MAGBY

Marbles was a big thing for a while. It would phase out and then phase back in. I can remember back in 1931 or 1932 when I was eleven or twelve, everybody was playing marbles, even the grown-ups were getting in with the kids. We had a better time than the kids do now. We created things to do.

We had a game we used to call "penny." It was like playing ball but we used a stick that was drove into the ground. You'd hit it with another stick about three- or four-foot long, not quite as long as a hoe handle, and catapult it into the air. While it was in the air, you would knock it and then run the bases. I haven't seen "penny" played in fifty years. I've seen men get out on a Sunday afternoon and play that with each other or with the kids.

DULAS L. MASSEY

We run a lot. I guess that's why I've lived a little longer. We played "stink base." We had one road that run down the middle of town. The stores were on each side of the road. Those kids would get out there in the evening and choose up sides. Some of us would be on one side of the road, then some would be on the other side. We had a ring we drawed in the dirt. That would be the base. You had to be in that base where someone could come and touch you and get you out. We were running all over the country. We also played marbles, Indians and Cowboys. Mostly we fished and hiked over the Ozark Mountains.

MAMIE H. MAYS

Jump rope, hide-and-seek, tag, drop the handkerchief. We couldn't dance but we would have "play parties." After my mama died and Daddy, who was seven years younger than my mother, was out hunting him another woman, my uncle from California was staying with us. We had an old Victrola record player. So while he was with us, he taught me how to waltz! [chuckling as she remembered]

EVELYN M. COONFIELD METCALF

The boys built wooden stilts, slingshots, and wood whistles. They would take an empty wooden threaded spool, rubber bands, matchsticks, and make cars. They played marbles, leap frog, and hide-and-seek. We walked on tin cans when we could find them. Everything was recycled in some fashion. We took a wagon wheel metal ring and rolled it around with a stick. Sometimes we would climb inside an old tire and have someone to roll us. After a while you learn how to keep your balance.

Jump rope was always a good game. We would jump together and say, "Miss *(first name)* went upstairs and kissed her fellow. How many kisses did she get?" (*Start counting: One two, three, etc., until she misses a jump, stop counting.*) A good jumper could get up into the big numbers. It was a game everyone got in and played.

Drop the handkerchief. We would form a circle with one leader. The leader would go behind you and all of us would be singing, "Who got it? Who got it?" Then pointing our finger at someone in the circle, we would sing, "Do you have it?" When the hankie was dropped, the person who had it drop behind them, would start singing, "I have it! I have it! Here I come . . ."

Button, button, who's got the button? Form a circle with one leader. Place your hands in back to form a cup. The leader would have a button or usually a small stone. The leader would go outside the circle as we would be saying, "Do you? Do you?" and repeat it until the stone had been dropped into somebody's hand. Then that person would say, "I got the stone, here I come!" You would run after the leader and try to catch her. If you didn't catch her, she got into your place. The game would start all over again and you were the leader.

I went shopping. The idea was to use all the ABC's in the game. I would say, "I went to town and bought an apple." The next girl would repeat, "I went to town and bought an apple and a banana." The third girl would repeat the apple, banana, and then say "coconut," etc. It was fun until you forget all that the things that were bought!

Paper dolls. We cut paper dolls out of the old Sears catalogs and dressed our dolls with clothes cut out of the catalog before the catalog went to the outhouse. When we could get newspaper, we made sailor hats, and cut paper dolls connected together in sequence.

Dreams were cheap. And sometimes dreams do come true, such as me writing this for you. I have always wanted to write children's stories, but never got the opportunity. The lightning bugs always kept me thinking and reading, trying to understand how they made that little light shine? Where do they go in the winter? How long do they live? Do they return to the same place every year?

ALMA POUNDS

One time Mother and Daddy went to town and left us at home. That was when I was old enough to take care of the kids. Me and a bunch of my friends got out into the corncrib and shelled a bunch of corn and took it to the yard and had a war and used slingshots. We strewed so much corn on the yard, which wasn't grassy but was full of trees. We knew that Mother and Daddy would get onto us for strewing all that corn over the yard, so before they came home we all got busy and cleaned up the yard. There wasn't a grain of corn left to see. Mother and Daddy bragged on us for cleaning up the yard. [laughter]

JOANNE RIFE

My clearest memory of childhood games on the playground was that while boys played with knives in a game of spearing near a target or played marbles, etc., the girls played "house." We picked up rocks, laid them out on the ground to form rooms, etc. Maybe put some coats down to form a bed and thus designate a bedroom. We then played we were doing the things you do in a home like cooking. We took grass leaves, long ones, and wiped them with mud to make bacon. We took a kind of brown crumbly seed and played like it was coffee or cocoa. Our parents would give us an old chicken house of some little building for a playhouse. We didn't have those bright plastic houses from Wal-Mart that are in the backyards along the streets in the subdivisions of the younger parents today.

PAUL ROBERSON

After Sunday School and church, we'd go home and eat dinner and everybody would gather at our house. Then we would go walking through the woods to the little ditches shaking vines. Squirrels would come out and we would run those squirrels up and down that ditch bank. We'd throw rocks but we never could hit 'em.

And then one time—I'll never forget this one—we had gone to church and everybody had gotten together, and we said, "Let's go steal the eggs out of the nest, take them down and boil them in a bucket." I don't know how many times you have eaten four or five eggs without any water to drink. But we hard boiled all those eggs, cracked 'em, and ate them. We got caught when we went home looking for something to drink. The adults found out what we had done and we got a whippin' for that.

We stole watermelons out of Mr. Sy Hall's patch. He appeared on our porch one Sunday afternoon and said, "Mr. Jess (my father), can I come and talk to you?" So he came in and said, "I think your boys have taken some of my watermelons out of my watermelon patch." Daddy asked him how he could tell, and he said, "Well, Douglas Davis told on them." Douglas was about six-foot eight or nine and had a big foot. Mr. Hall said, "When I went out there and looked at that footprint, I knew who that was." So we had to chop cotton for a day with no pay from Mr. Hall.

ARWILDA WHITESIDE

"Heaven & Hell" suppers at church and school. The men in church would kill a cow. Dad would take the leftover bones, head, and feet and put them in a big pot, put in lots of water, and make hot chili. The soup was "hell" because of all the hot pepper they put in. When you bought your ticket, it was either for heaven (ice cream and cake) or for hell. Most of the folks "wanted to go to Hell" so they could get that chili!

CHARLES WHITFIELD

Hunting rabbits, skinning 'possoms and 'coons. We'd go to church and sometimes us boys would hunt a lot of the night. Back then, you could get twenty-five cents and sometimes fifty cents for a rabbit. You could catch an ol' 'possom, skin 'im, and cut his tail off and the head and feet and legs and take him to Hot Springs and sell 'im. Or a 'coon, either one. You'd dry the fur, stretch it on a board, hang it up until it dries, and take it to a place where they sell hides, you know. A 'coon hide, you'd rip it wide open and skin 'em, put it on a wide board, stretch it, and put tacks all around and dry it. When it got dry, you'd take it down and sell it. Sometimes you could sell it for two or three dollars. And a mink hide, they's up around twenty dollars, you know. Minks are hard to catch.

Cutting hair for free. A lot of times on weekends I would spend half a day cutting hair as fast as I could cut. Never did get a penny for it. Now it's eight dollars.

[Author's note: In 2010, make that twelve dollars at Unique Barbershop in Arkadelphia!]

Chapter 17

Courtship and Marriage

FRED NORMAN BLANKENSHIP
Well, I tell you. It was more of a social group thing. I didn't have a serious date until I was nineteen or more. I rode a saddle horse four miles to go courting on Saturday evening to Elsie, the girl I later married. On Saturday night, somebody would give a musical. We would meet at somebody's house and we'd play games like spin the bottle and tap two—we'd get in a circle to get boys and girls together. Dating was more of a social thing that kind of grew into a courting situation, getting kinda struck on a girl. When I got ready to go see Elsie, I asked my mother if I looked all right. She would say, "If you do as good as you look, you'll be fine." Sometimes even today when Elsie and I get ready to go somewhere, I ask her if I look all right.

R. L. "BILL" CARTER
I didn't date much until I went into the navy. I'd go to play parties there at people's houses. We'd play games like "fishin' for love," and all that. The girls would hold their hands out there with their eyes shut and you'd stick you finger in their hand and they'd grab it. Then you would get to take them for a walk. [chuckling] I used to go to old country dances. They'd have a fiddle and a guitar. And they'd move what little furniture there was—there wasn't very much back in them days. They'd get over in a corner and play and we'd dance.

MARY LEONA CARTILLAR
Russell [Cartillar, her future husband] *lived at Tilton and I lived in the country near Hickory Ridge.* He walked to my house, which was about ten miles away. I met him at a dance. I knew him before, but I didn't want to date him.

Why not? Well, he had the name of being kinda rough, you know. Daddy wanted me to date another boy. I wouldn't do it. It was Gene. He wasn't as bad as Russell, but still I didn't like him, either, as well. I dated him one time and I didn't date him no more. But anyway, Daddy always said, "Date Gene, 'cause Gene likes horses." And Russell didn't. Of course, everybody thought

I was gonna get in trouble right there, but I didn't, 'cause I knew how to take care of myself. He wouldn't have kept walking up to see me, but I was kind of a challenge to him, you know, 'cause he couldn't do nothing with me. He kept coming up there and so finally we got married.

Daddy was the one that told us about sex. He had all girls and he seen that there was nothing happened to any of us—I tell you! He was a real—in a way—a hard man, but he seen there was nothing happened to us. But he told us about life, I tell you. He would tell us that if we dated—went out with a boy—he would say, "Now they'll all be crazy about you, but don't believe them!"

I mean, he told us facts, and it stuck with me all my life—I'm telling you. Because I knew he knew what he was talking about. Now, most men wouldn't do that. But, he did, 'cause he said, "Now your ma"—he called Mama 'ma' to us, you know—"Now your ma won't tell you." And, that was fine, because nothing ever happened to one of us girls either 'cause we knew.

I had a half-sister. She was about two years younger than me. She's dead now. She was a tough character. Would fight. Them boys never tackled her, I can tell you, because she would land them in a road ditch. Boy, she was something else! I tell you, we survived—but I'm telling you, it was hard.

How did Russell propose? Well, he just asked me. He didn't get down on his knees or anything like that—he just asked me. At the time, I didn't think I was gonna get married. But one day, as time went on, he got an old car, came up to the house, and brought a man with him. I asked, "Where we going?" And he said, "Well, we're going to get married." And we went to a justice of the peace, one of Russell's neighbors. We got married out in kind of a pasture—like, on the grass, you know. There was no ring or anything 'cause there just was no money. Ann Webb [Mary's daughter, who was listening to the interview], said, "Daddy would always tease us and tell us you got married on a cow-pile." Mary's reply: "Yeah, that's right! We got married in a cow pasture!"

WILLIAM EDWARD DELAMAR

How did you meet your wife? I'm only eleven months older than my wife. She's a full-blooded Methodist. We were maybe fifteen years old. She's still Methodist and I'm still Baptist. Some people say Methodists and Baptists can't get along. I don't know about that but we're still together.

[Author's note: At the time of this interview, Mr. and Mrs. Delamar had been married seventy years.]

GENEVA KING EMERSON

Steve and I met at church. When we went on a date, it was to a church youth meeting with other young folks. Decent girls didn't run around alone with boys. Young people didn't touch, no matter how enamored they were of each other. Even married folks abstained from any public demonstration of affection.

EVELYN LANGLEY

Somebody would give a party every Wednesday night and Saturday night. We'd have a dating game. I have forgot how we got our partners. I believe we all had numbers. Somebody would call the boy's number and he would come in and got whoever that girl was and they went walkin'. We'd also play spin the bottle or "spin the plate" (or a bucket lid!).

Where did you meet your husband? We went in a wagon to my aunt's house. They lived close to Guy. Farris Wilburn Langley, who later became my husband, was their next-door neighbor. We'd all meet out there in the field and play ball. So we got acquainted when we was just kids. And when we moved to Guy, first thing I knowed, here he come! [laughing] I was thirteen then and we started datin' and we dated until I was eighteen, when we married. We were married for forty-eight years when he died of cancer in 1986.

What memories do you have of the marriage? Farris asked for me a day or two before we were married on a Saturday night. I can just see him. Daddy was sitting on the porch and Mama was standing at the door. Farris walked up there and said, "Me and Evelyn's planning on getting married on this weekend. I'd like to know if youall have any objections." They said, "No."

So he came back on Saturday evening and got me. He borrowed his brother-in-law's ton-and-a-half truck and come up to my house.

Farris lived way down towards Greenbrier and I lived up the other way towards Guy. Mama ordered me a dress from the catalog and that was my wedding dress. It wasn't much but I thought I was dressed up. He came and got me and my two sisters and we went to Greenbrier to a preacher's house, a Mr. Wilson. And I never will know how come him to do it, because Farris told me that he had told the preacher that we was coming to get married. But when we drove up, he came out there and asked us to come in. And Farris said we'd come to get married if he wasn't busy.

"Well, just let me go get my Bible," he said. So he went back into his house, got his Bible, put on a coat and tie, come out there and stood beside the truck—we never did get out. [cackling with laughter] We sat in that truck and got married. My two sisters was the witnesses. When it was over with, Mr. Wilson went into the house and we left, and went to church in Guy.

It was during a revival. We went on to church and my close friends knew I was gettin' married that night. Well, they just swarmed me to congratulate me. When church was over, we let the rest of them know that we had gotten married.

MAMIE H. MAYS

When FDR became president, he started the CCC camps and the WPA down here. He gave the people a boost and something to do. His programs made jobs. The CCC 746 Company was at Lono, where I found my true love, Hartsville Regan Mayes, from Johnson County. It was love at first sight. I asked him, "How come you want to marry me?" He said, "I knew you were a good girl. Boys talk, too!" We were married in 1936 and he died in 1989. We were married for fifty-three years. Two children, three grandchildren, two great-grandsons.

WILLIE MORRIS

I remember this one girl I was dating. She lived up the bayou from us and I had to go through a cemetery. When I got ready to come back that night, it started raining. I was wearing some old corduroy pants and was getting pretty close to the cemetery, which scared me anyway because I had been told so many scary stories and I thought someone was following me! I didn't realize until I got out of the cemetery that the sound I was hearing wasn't somebody following me but was those corduroy pants legs rubbing together—*whisk! whisk! whisk!* The faster I would go, it sounded like somebody was about to catch me! When I got to the end of the cemetery and could slow down, I could hear the sound going a lot slower. That sound would have had me runnin' all night!

EVA SMOKE WELLS

We mostly went in groups. A family moved in a house on our road with several boys and girls and we all went to church together. We walked both ways and when church was over, we were strung out a half a mile on the highway.

Where did you meet your spouse? He had been in a CCC camp in Idaho. We met at school during a ball game. It was about 1939. I was just fifteen. He was just different from the boys I was used to. He'd been places I hadn't been.

[chuckling] But we had forty-seven years of marriage before I lost him in 1987.

ARWILDA WHITESIDE
Cleovis and I met at the fork in the road. I was nine years old and he saw me and told them he was going to marry [me] and he did when I was thirteen years old on July 24, 1939. Sixty-nine years later, twelve children and we don't know how many grandchildren. A lot! Thank God, we still love each other. We celebrated our sixty-ninth anniversary on July 24, 2008.

CHARLES WHITFIELD
Grace and I didn't have any kids to start with. We had only one boy. We growed up right close together. Seemed like I was about twenty-three when we married. She must have been about nineteen or twenty. I've heard her laugh about us waiting to get grown. We went up to Billy Shuffield's house and got married. George Cook and Billy Thomas carried us right up here. We didn't have no car, so we went afoot. They carried us up there and we came on back home. We didn't have any trouble a'tall. She was helping me when she was able. She was sick a lot. She'd help me hoe cotton and pick cotton some. In her later days, she didn't try that much. She always had somethin' cooked for me.

FLOYE WINGFIELD
Mama didn't allow us to date until we were sixteen, but that didn't keep me from lookin'. There was no transportation for the young people. I remember on Sunday nights we always went down to Uncle Bob and Aunt Penny Green's. They had a Victrola [record player]. Uncle Bob would run the Victrola while the kids danced. He just tapped his foot and had a good time.

Where did you meet your spouse? I don't have a husband, I've never had one! I'm the community spinster. You can take me as I am or you can just leave. I betcha anything I'll keep on ticking. [laughing]

What do you want on your tombstone? I do not want "Miss" on it because I have not missed anything. [still laughing] That's the truth! That's the truth!

[Author's note: Before the irrepressible Floye Wingfield died in 2010, she arranged for the funeral procession to stop at the entrance of the cemetery so her mourners—she would prefer to call them celebrants—followed the hearse and a New Orleans–type band playing full blast to the grave site.]

Chapter 18

Medical Care

CELIA ALMYRA GRAHAM ACREY
My daddy was a medical doctor but quit practicing after my mother died because nobody had the money to pay him off. So he had to do something to make a living. He just quit and bought the farm and started farming and raising stuff to sell.

Mama and a bunch of her family died with flu in about 1918. Grandma's two sons died five days apart. The problem was they couldn't get no medicine, they couldn't get nothin'. Even my dad couldn't get it. He wouldn't get nothin' for delivering a baby. If they had something to give him, like chickens, they would. But a lot of them didn't have them. He just couldn't make nuthin'. They were some very hard days.

KATHRYN BAJOREK
My family prayed and fasted for our medical needs. I was at the doctor once before I married because I stepped on a board and got a nail in my foot. Coal oil didn't help. After my dad had torn down a building, he told us, "Don't you girls go down here and be jumping the boards." But that's what I did. My dad prayed and prayed and my mother said, "If you don't take her to the doctor, she's going to lose her foot!" So he took me to the doctor. If my family had not had faith, we would have probably all died.

BESSIE BLACKNALL
"Doctor Bob," as we called him, lived in Gum Springs and had a horse and buggy. He didn't have a stethoscope, so when he wanted to find out what was wrong with us, he'd put his big ears next to our heart and listen to our chest. Then he would give us a laxative that would make us use the toilet. You couldn't do anything until that medicine got through. At that time, they believed in purging you. With the faith of the Lord and going to the bathroom that much, you probably would have forgotten you were sick in the first place. He'd deliver babies, too. All the children around where we lived except me were delivered by midwives.

R. L. "BILL" CARTER

Doctor pays for operation. My mother had a bunch of varicose veins on her leg and had a sore about as big around as my fist on there and she had to bear that for several years. Dr. Valentine Pardo, the old Cuban doctor, wanted them to have it operated on but they didn't have the money. So one time he said, "Take your mother to Little Rock and get that operated on and I'll pay for it." And he did! Well, Dad couldn't stand to owe him any money. He talked to him three or four times about it and the doctor the same as told him, "You don't owe me nuthin'." But Dad was never satisfied so he sold thirty acres of the back side of the old farm to pay for it. Dr. Pardo would come when we would call, regardless of how the weather was.

There was one doctor in Blackton (Monroe County) that delivered Mom's first baby. She was having a hard time. Dr. Bradley rode a horse about eight miles from Blackton to Rich. He told Daddy he had forgotten to bring his forceps. So he rode his horse back to Blackton on the other side of Monroe County to get his forceps. [When he returned] he told Dad to hold Mama down and put a spoon in her mouth so she wouldn't swaller her tongue. And that's how they delivered that baby. All the rest of [the brothers and sisters] were delivered by a midwife, all born at home.

VERNON L. GLENN

The time a wagon wheel ran over my neck. When I was about five (or four) all the other children were at school, I went with my mom and dad to cut some firewood. When the wagon was full, we were bringing it back to the house. My mom had walked ahead to open the gate for Dad, and I guess I was twisting around. I fell off that wagon, and when I did, my feet flopped over and my face went into the ground. That back wheel (loaded with wood) went right across my neck. I know it was a drought, because I remember my mouth was full of dust. Anyway, it scared them—they grabbed me and took me to bed and got me quiet. While my mom was doing that, Dad unloaded the wood so they could head for town to see the doctor.

We headed for town and took us a half an hour or more to get to the doctor. He was one of these that kinda walked around town talking to people. By that time, I was up walking around. After he checked—had me move my neck around—he said, "Go buy the boy some candy and he'll be okay." My dad gave me a nickel, and back then you could buy Baby Ruths for a penny, so I had me five Baby Ruth candy bars. I said that was the best medicine I ever had. It's just one of those things where you wonder how you got through life.

WILLIS MAGBY
We had a bunch of sickness when I was seven years old and there were five of us children at the time. I was next to the oldest in the family. In fact, at one time, the whole family was down with typhoid fever except me and my dad. There would be four in bed at one time. At times, Dad had to be away a while—an hour or two—and I had to take care of all their needs while he was gone. My youngest brother was a little less than two years old. He was just barely walking. He had bronchial pneumonia at the same time. They laid him out for dead but then they worked on him and he come to.

We had a local doctor who was about four miles from where we lived here in Point Cedar. He was a good doctor but he didn't recognize that it was typhoid fever we had because there wasn't much typhoid fever at that time. So he was treating us as if we had the flu (influenza). They wouldn't give us hardly anything to eat, just a little thin soup, you know. You've heard about the old folk remedy, "Feed a fever and starve a cold"? He was doing just the opposite—he was starving a fever. If you don't have any nourishment at all, you can't make any progress. But at that time, doctors thought that was the thing to do—we're talking about more than eighty years ago.

EVA SMOKE WELLS
Sewed up without antiseptic. The first time I ever remember being in a doctor's office was when I was eleven years old. About 1936, on a Saturday, we were playing marbles. Everybody said, "Let's go get a pear." I beat all of them around there. I was first to climb the tree. My feet were muddy. I climbed the tree but I fell down. As I came down, my arm caught on a limb. When they took me to the family doctor, he was "indisposed." We got another doctor who laid me on the table in his office—on my stomach—and sewed that up with no antiseptic whatever. He even cut off a chunk with his scissors that wouldn't fit right.

Why did he do that? Because he didn't think he would get paid for it! He thought my daddy was a poor farmer. That's just one of the times. Today, I have to go out to the hospital every two weeks to get a blood test. Everytime I go out there I see a tombstone with that doctor's name on it and I wonder if it's him! [laughing]

CHARLES WHITFIELD
When my son Pete was born in 1931, that cost twenty dollars. They didn't go to the hospital to have babies. They would have them at home. I owed Dr.

Hodges twenty dollars for delivering a baby. That's equal to two hundred dollars now. That fall, when I would do my crop and pay for my groceries, I was broke. I never will forget this. I went down to Dr. Hodges—he had a heart in 'im. I said, "Doctor, I don't aim to beat you. I will pay you sometime, but I don't have a dime. If I can find work to do, I'll pay you."

"You want to work?" he asked.

"If I can get a job," I said.

"Come Monday morning, you got one." And I went to work for one dollar a day. And I hadn't been there but for a few minutes when Arlie Moore, who run a garage over there, he come. He was going to work out his doctor bill. That old doctor let us fiddle and work to pay our doctor bills.

Chapter 19

Home Remedies

DOROTHY COX COSTON ASHLEY

Stopping a cough. Two drops of kerosene in a teaspoon of sugar would stop a cough—and still will! Camphor was used a lot. Iodine and axle grease were used for cuts and abrasions. We didn't treat bruises. They got well on their own. Soot from the fire place was used to stop bleeding. Everybody had a jar of Vick's Salve and a bottle of alcohol.

Asophidita bags. My sister and I were given a "round of quinine" each spring, consisting of hideous, bitter, yellow pills that had to be cut in half for children. Mother buried them in oatmeal but they were still bitter. Some children wore smelly asophidita bags around their necks to ward off colds, flu, and other winter-related illnesses. We used to drink sassafras tea every spring for something, but I'm not sure what it was supposed to do.

[Author's note: An asophidita bag was a folk remedy most commonly found in the Appalachian region in the eighteenth or nineteenth century. Basically, it was a pack of pungent herbs, often including ginsing and yellow root—the exact ingredients varied by practitioner. The vapors were supposed to ward off colds or other diseases.]

The itch. In the winter months people suffered from a skin condition that we just simply called "the Itch," which was probably an allergic skin reaction to some little bug. I was allergic to some of the fabrics we were using that was aggravated by scratching. Sulfur mixed with grease was used to successfully treat it. The mix smelled terrible but people used it.

BESSIE BLACKNALL

Sugar used to stop bleeding. When you started bleeding after a cut, they didn't know then you could stop the bleeding by putting pressure on the wound. What they would do was to put a bunch of sugar on it. By the time the sugar got in there, it would cause the blood to clot.

Chest colds, etc. We used pine tar. They would take the needles and mix it together with mullen and sometimes they would get holly roots. You could boil it all together on the stove or do it separately. Then they would put sugar in it and make a "toddy," which meant they would put plain ol' rot-gut whiskey in it. Then they would sprinkle a little sugar in it. I'm told it tasted pretty good when you get through with it!

Pinkeye (conjunctivitis) *or a sty on the eye.* Rub urine on a baby's diaper on the eye and, by the grace of God, it would heal. If God didn't have anything else to do, it would heal. Now they give them eye drops for the pinkeye.

The thrush. When a baby would have the thrush, their mouths and sometimes their lips would turn real white on the inside, or little festering bumps would appear on the tongue. Their mouths would be so sore that they wouldn't suck on a bottle or a breast. There were about two men that lived in the community who could treat it. They would take these babies to one of these men, who would open that baby's mouth and blow in it. And that would heal the thrush. Babies still have the thrush today but doctors swab their mouths with microstatin, an antibiotic, that gets rid of it.

Kidney infections. They would go out and dig a peach tree root, boil it on the stove, and give the liquid from boiling it to drink. The main truth of this was I really do believe that they had not been drinking enough water.

Getting rid of worms. If babies or children didn't sleep good, they would ball up on their knees and sleep on their arms and head, people would say that they were "wormy." Mama would either give them a piece of sugar with a drop or two of turpentine or give them a spoonful of water mixed with some bluing—the kind we use today to bleach clothes—and the worm would come out through their feces.

We would follow behind them when they went to the bathroom and you would see the worms, maybe four or five inches long. I remember when Mama gave a child some turpentine on sugar. When he got to the bathroom, his feces were full of worms! [laughing]

MARY LEONA CARTILLAR

My grandma was what they call a "midwife," so she doctored us with whatever she could get. I loved pickles. I got sick and she said, "Now, don't y'all let her have none of them pickles." I had chills—do you know what chills are? I don't hear about them anymore. My sister, Hazel, brought me the pickles, and I don't reckon they hurt me a bit!

She used a lot of quinine. And she also used a lot of castor oil. I hate that stuff! A baby died in our community, they said with diphtheria. Daddy said—now this is not no doctor—he said that the baby died from "cats." They had a lot of cats around—I was on the bus and . . . I told you I was hardheaded. They was out in the cemetery burying that baby. My daddy was out there. They didn't have no funeral because diphtheria is catching. The bus driver stopped and let me off. I was out there with all those men. I don't know if they put that baby in a coffin or not, but I've got a feeling they might have put it in a little wooden box.

I didn't see the baby, but anyway, Daddy took me home—him and his brother. Daddy said, "I don't know what to do, she's come out there and will probably catch what that baby had," and so he said, "We'll give her a dose of castor oil."

"I'm not gonna take it!" I cried.

"Yes, you are!" he said.

They got me up in a chair and then Daddy said, "Open your mouth." They opened my mouth for me and I spit it right straight out every time. I'll never forget what he said: "Oh, hell, you couldn't get it down her."

DOYLE L. COLLINS

Stone bruises. One time, I developed a bad stone bruise on my heel, probably caused by going barefooted in the summertime. With that constant pounding, an infection developed, which was a common occurrence. My ankle swelled up and got real painful and sore. My folk kept examining that. They thought it was ready for lancing. My daddy got out his strop and straight razor, sharpened it up good, and my mother supplied a couple of long needles. They laid me down on a cot and lanced that stone bruise. My heel got well.

WILLIAM EDWARD DELAMAR

Chest colds. Most old people today used home remedies when they were young. With a cold in your chest, my mama would use beef tallow. When we would kill a cow, we got the fat from the meat, and boiled it down to grease. Once it gets cold, the tallow is almost like a bar of soap. Then she would rub it on our chests with a wool rag.

Mama also used mullen, which is a plant. She'd get some green pine needles, boil 'em in water with the mullen and make a tea. We would drink it as hot as we could drink it. Then Mama would put us in bed and put the covers on. We'd break out in a sweat like you would if it was summer and someone would put a blanket on you. The sweat would pull out the cold.

Sweating out fever. We would use hogs' hooves. After killing a hog, we would take off the hog's foot, put it in water, and turn the water up to a boil. Then we would drink it as hot as we could stand it. Boiling the foot changes the color of the water. It would sweat the fever out. If it was good then, it's still good now.

GENEVA KING EMERSON

Worms, chigger bites, tick bites, snakebites. I was delivered by a midwife, given turpentine and sugar for worms, "doctored" my chigger and tick bites and scratches with turpentine. Snakebites were soaked in kerosene. We had lots of snakes about the farm, even in our yard and garden. It's a wonder we seldom suffered bites.

My sister was bitten by a snake. She sat with her foot soaking in a basin of kerosene with careful watching for blistering by Dad. Probably, the snake was not one of the really poisonous ones, though there was a lot of swelling, which Dad also carefully watched. My sister didn't experience a great amount of stomach sickness and her leg gradually became normal over several days.

VERNON L. GLENN

Measles. A really true story. During the school year, measles came along. Word was that you had to stay in bed for three days. If you went outside, you'd go blind. So we put things on the windows and my sister and I stayed in there three days. I tell everybody it worked because we didn't go blind.

GEORGIA HEARN HILL

When a child was sick, my mother and Aunt Alma would work together. They would call the oldest mothers in the community to come in and get the fever down, or whatever. They knew what to do—home remedies.

Cow-chip tea. One of the home remedies was cow chips and mullen tea. My aunt Mary Jane, when children would have a really bad cough, would get a dried-out cow chip, wrap it up along with several mullen leaves and several chips of pine bark in a white, clean rag.

Then she would boil it an hour or two down to a tea, and sweeten it with sorghum molasses. It would be like ice tea. [sustained laughter] We would then give it to the children while it was hot, maybe adding a little lemon juice to give it a little "whing." [more laughter] Well, we got well, we got well! There weren't any chemicals in it like there is today. It was just grass and corn and whatever Daddy and them fed the cows.

My grandmother, Mama Georgia, was a midwife so she brought forth many, many, many babies in Gum Springs, Curtis, all around down there. They took care of the babies. I was born in my grandmama's house in Gum Springs.

KENNETH GUY LACY

To stop severe bleeding. Mom would sometimes melt beef tallow and put some in the wound. That certainly sealed it! In lower elementary school, I fell, hitting my head on the corner of a homemade bench, cutting a big gash in my head. Some of the older boys carried me home. Mom used the tallow to stop the bleeding. The scar caused me to be turned down by the army on my first examination, but because they were needing more soldiers during World War II, I passed the next exam.

Malaria. One summer I had a case of malaria for several weeks. Mama brewed of peach-tree-leaf tea for me to drink. It was certainly a bitter dose to swallow! They finally got a store-bought bottle of "666" that did taste much better.

Colds. She also would make a poultice of onions—sometimes raw and sometimes cooked—to put around our necks for colds. Vick's Salve was also for colds. She would put some on a flannel cloth, warm it, and put it on our chests.

Sores or boils. Mama made a poultice of a raw Irish potato to put on sores or boils that needed to open and run out the pus. That might have to stay on a day or so.

MAMIE H. MAYS

Every community had a lady who could give home remedies. My grandmother, Bettie Guest, was one of those women. Everybody called her Aunt Bettie. She could treat horses and people, too. If we had fever, my mama would also send for her and she would come and fix some cough syrup.

Sore throat. For cough syrup or sore throat, she used honey and whiskey. Everybody kept a little whiskey for medical reasons. Old white moonshine, I guess. But they would put just a little bit of it in there and you would take it. She would take flannel rags and smear Vick's Salve on our chests and we'd have to wear it to school. Or sometimes they would bathe our faces and arms in cold water. If we had fever, they'd call a doctor in the Willow community near Malvern.

The itch. The worst thing we ever caught after [World War I] was over was when some of my daddy's nephews come by to see us and they brought us the "itch" that we all got. To treat us, Mama would get pine tar and sulfur and rub it all over us and got on our underwear. It smelled awful. But everybody just laughed and grinned.

EVELYN M. COONFIELD METCALF

Sore throats, etc. Cod liver oil was in all the kitchens. A teaspoon of turpentine and sugar added was used for sore throats and chest colds. Cuts and scrapes were treated with iodine.

Walking pneumonia. We would cook onions until done and hot. Then we would place them in a towel as hot as you could handle. Then a larger towel or material was wrapped and tied around your chest. The smell wasn't too bad. This was used for a chest cold or "walking pneumonia," as we call it today. We could work or at least be up and not in bed. To loosen up the phlegm in your chest, the blankets were pulled up tight.

Treating boils. Boils were a pain. My grandmother Coonfield would take a needle and punch at least two holes in the middle of the boil head and then place the egg white silk—the white strip inside an eggshell that looks like silk—on the boil and cover it with a bandage. If that didn't work, bacon or a piece of fat was placed on the head of the boil.

Carbuncles. Once I had a carbuncle on my chin. When it got yellow, my grandmother would take a string and make a loop around the boil with the string and twist it till the core would pop out and all the pus behind it would come out. It did hurt a little.

Skunk oil. A skunk was caught, killed, cleaned, and baked in the oven. A skunk is real fat and greasy like a fat hen. The broth was kept in a jar and heated as needed. It was used like chicken broth but, like medicine, just a tablespoon or two at a time. It was used for a real bad sore throat like strep throat.

Mustard plaster. Used on children to loosen up the phlegm in the child's chest. It was a little like Vick's Salve today. To make it, you would boil mustard and then cool it, make a paste with flour. Place on our chest and back, rubbing it with warm hands and hot towels to keep the plaster working to loosen the lungs. You were wrapped with towels or material real tight to keep the heat in. Children could run and play in the house. The plaster was peeled off when it got cold and your skin would be a little green.

WILLIE MORRIS

Stomach problems, constipation. The plantation had a doctor on the premises. But most of the time, we had some older people in my family that was good at making medicine. One thing we used was May apple roots. We would make a tea out of it. And if you had a stomach problem or something like that—or constipation—that would take care of it.

WILLIAM PIERCY

Tetanus. Mama doctored it by dipping our toe in kerosene and wrapping a rag around it. We would tear up sheets and use them for bandages. No Band-Aids back then. We never had infections. I ripped my leg open on a barbed-wire fence. Worried about tetanus, Mother soaked a rag in kerosene, laid it on the cut, and wrapped it. And it worked.

Boils. Mother used Irish potatoes as a poultice. We had a lot of boils when we were kids, probably for lack of good nutrition. She would grate an Irish potato to get out the juice, lay a rag with potatoes on the boil to make it drain. Didn't have to have it cut open.

Stone bruises. One time while Mother had been hoeing all day, she bruised her hand so bad she couldn't sleep. A neighbor told her to take a jar of her pickled beets, mash them up, fill her hand completely with the beets, wrap it up in a sheet, and let it stay there all night. When she woke up the next morning, the bruise was totally gone.

Purging our systems: Dig up sassafras roots to make tea. Mother would use it to purge—clean out our systems every spring.

ALMA POUNDS

Sore throats. Get a spoon of sugar, put a drop or two of turpentine in it, and put it in your mouth to melt and go down your throat.

Worms. Mother would wash egg shells and dry them in the oven. She would roll them out to just as fine as she could get them and mix them with sugar. She imagined that would be for worms. If we ever had worms, I didn't know it, but she was going to make sure we didn't!

Treating cuts. One time I split my ankle I think to the bone or at least to the tendon. I know it was a real deep cut. I went in howling, of course, and Mother came out to see what was wrong. When she saw the cut, she ran back into the house and got a handful of salt and put in on. That cured the cut. The ankle healed without any trouble.

UNION H. STOUDAMIRE

If we got sick, the doctor wouldn't come out to the house. The old folks gathered weeds and herbs from the field. They would give it to us like in the fall of the year. We made our own medicine and cured them. I think if we were still doing that, there would be more folks living and not so many folks dying as there is now.

My cousin had a great big medical book she ordered. It had every kind of medical weed, everything in there a doctor would need to know. And she would get a big sack and put it on her shoulder, get an axe and a hoe, and go down there in the woods. With this book, she knew every weed, every bush down in there. She would come back with a sack on her shoulder full of this thing and that thing—this kind of weed and this kind of bush and this kind of limb and this kind of leaf, and she would make our medicine. We would take it and we would get well! We didn't hardly get sick back along then no way though.

Cow-chip tea. I have taken that! Folks would go out to where the cows lay down at night. The cow chips had already gotten dry and hard-crust like. They would dig up and bring it home, put it in a pan, and boil it. I cried, because I didn't want to take it, but my aunti made me take it anyway. We got well.

Horehound. The old folks would [use] a weed was called wild horehound. It was good for children with stomachache. They would give us dry cow-chip tea for meals. The nasty castor oil for a child with worms. The old people didn't have money to go to a doctor and some just didn't know about doctors back then.

ARWILDA WHITESIDE

Hog hoof tea. After you cut off their toes, let them dry then put them in the oven and bake them. Then pour hot water on them and make tea. Used for bad colds and pneumonia.

Cow-chip tea. If you want to know the truth about cow-chip tea, my daddy had told my mama that if anybody ever gave him some cow-chip tea he was going to kill 'em! But once my daddy got deathly sick and the doctor told us he was not going to live. So my mother made him some cow-chip tea. He got up. And my uncle, who was walking around all day, fell dead. But my daddy lived thirty, forty, or fifty years after that. But Mama never told him that she gave him some cow-chip tea. That's the truth!

Sheep-balls tea. When they would "dump," little balls fell out. We'd dry them, boil them, and use it for tea. That and cow chips were used for pneumonia or bad colds.

Mullen. They would get the mullen leaves. If you had a swelling or have pus in your feet, anything like that, they would lay some mullen on them. If you had a fever, they would let peach tree leaves soaked in vinegar. I had pneumonia so they would lay the peach-tree leaves and vinegar on me.

Horehound. That was a weed we would cook down into a syrup to use by the spoonful for sore throats.

Quinine. My daddy also believed in quinine that was used when we would have fever. If you had bad coughs and fever, they would put some quinine on you. My daddy would go to a drug store and buy these little empty capsules. We would take quinine in capsules if you had bad coughs or fever. Daddy would go to the drugstore and buy these empty capsules and fill them with quinine and use CC pills you could buy over the counter.

Tallow. If you had pneumonia or bad colds, that's why you would rub down your chest or back with tallow—beef fat—and cover it with a piece of flannel. This would help break the fever.

Black Draught is a laxative that's still available. It's used as a purge.

Corn-shuck tea was used for pneumonia, colds, clean out kidneys (*have faith in God!*).

Lime as a mouthwash to keep your teeth from decaying. We would use a half a cup of lime, pour water on it. After so many days, poured out the water, put some more in. Let it stand until the water got clear then shook it up. After the lime would go to the bottom, pour it off, put in more water, shake it up, lime goes to bottom of the jar, pour it off every day for seven days until the water is clear. Then you could use it.

Chapter 20

The Drought of 1930, Other Weather Stories

DOROTHY COX COSTON ASHLEY

We had no daily or weekly weather reports so we went by "what the old folks said." They had such sayings as, "Thunder in February, frost in April," "Rain on the first day of the month and it will rain fifteen days in that month." The first twelve days after Christmas were supposed to predict what the weather for the coming twelve months would be like. December 26 was supposed to predict January weather, December 27 was supposed to predict February weather, etc. If it rained on these days, the month would be wet. If it was a sunny day, the month would be sunny. If it was a colder than usual day, the month would be cooler than usual, etc.

Other little sayings we went by. "Morning red and evening gray will send a shepard on his way. Morning gray and evening red will send a shepherd wet to bed." An early Easter always predicted a late spring. And of course everybody knew that geese flying south in the fall predicted a cold spell and geese flying north in the spring predicted warmer weather. We watched the blackbirds gathering into flying flocks in the fall. And as their flocks gained in size, we knew a cold spell was coming. And we always knew there would be a cold spell when the blackberries bloomed. We called it "Blackberry Winter."

KATHRYN BAJOREK

One year, during a bad drought, we had no money. At that time my dad had bought an old truck. He said to my mother, "Prissy, we are going to load all these kids in the truck and we're gonna drive to Pine Bluff. There is a spiritual man coming to church who believes in prayer. I'm gonna fast and pray that we're going to go down there. I have one hundred dollars and I'm going to give money in the offering. It's all we have. I'm gonna pray and believe that God will send us some rain."

We went down there. Daddy never went up to the front, but when that man asked if anybody had any spiritual prayers, my dad stood up with a hundred other people, I guess, and he started praying. When they passed the offering plate, my dad gave one hundred dollars and my mother said, "You know, there's so many things we need that we could use this one hundred dollars."

Dad said, "If two or three are gathered together and pray and ask and believe, it will happen."

That night, when we were driving home, we were in back of the truck. The windows was knocked out of the truck. My dad always kept old worn-out cotton sacks that had tar on the bottom that he would hang over the door so he could slam the door and hold the sack in place so rain wouldn't come into the truck. On the way back home, it started raining. My dad got out of the truck and stood there in the rain and just prayed! He finally got all those cotton sacks with tar over the windows so the rain wouldn't come in. My sister and I were in the back and I remember crawling up into them old cotton sacks so we wouldn't get just drenched. That year we had some of the best crops we'd ever had while everybody else's was burning up.

LOUISE J. BRANT

There were good years and bad years with crops. The flood of 1927 got into our house so we had to move out. It was just like it is today. We just had to depend on the rain. That was the only moisture we had. We didn't have irrigation, of course, which means that if it didn't get rain, we would just lose the crop like you do now. We had years of too much rain and years of not enough, the way it's always been.

FAYE HARGIS ANDERSON BYRD

In 1930 after Bill and me married we had a drought. We had made one bale of cotton and a half a load of corn, and we had to borrow money to buy fertilizer and it took all that to pay for the fertilizer 'cause cotton was cheap then. I'll tell you what we done. Mr. and Mrs. Anderson had moved down here and rented a place and they paid thirty-five dollars a month for rent for a house down here. They moved back in part of the house with us and they helped feed us through the winter. Bill went to Missouri and he went everywhere and you couldn't get a job; there wasn't nothing to get.

You could get apples, tangerines, oranges, and things like that once a month. It was sort of like these commodities. Then I've seen the time when we didn't have milk, and Aunt Mary Moser lived not too far from us and Bill would go once a day to get a gallon of buttermilk. I'll tell you that was good thick buttermilk, too! And he'd walk and go get that.

[Author's note: Excerpt from "Down Memory Lane." Used with permission.]

TENNESSEE "TENNIE" LAWHON DOVER

We had a drought. Lord, have mercy; I went through them. I'll never forget it. It was hard times but there wasn't anybody but Harvey and me, and I'll tell you there was families that lived right close to us that had a bunch of kids. One family lived up this way and the other back this a'way from us. And I'm telling you right now, they'd have both starved to death if it hadn't been for Harvey and me. When we'd get done eating of a night we'd always have something to feed the dogs and Harvey said, "Don't throw away one thing. Put it away; think about them little old young'uns that hadn't got nothing to eat."

[Author's note: Excerpt from "Down Memory Lane." Used with permission.]

BEULAH LEE MCLEOD EVANS

One of the times we used our storm cellar was when Daddy came in from work one afternoon. It was clear, just as clear as a whistle. The wind had blowed all day. And there might have been a little bit of cloud in the Southwest, I don't know. But Daddy came in early from the field, and he said, "Viola [Mrs. Evans], fix supper. Let's eat early. It's gonna storm." And so she did. He said, "Now put the fire out in the stove." So they put the fire out in the stove, and they covered it with ashes to smother it down. We was eating supper and Daddy said, "Let's go."

It started lightning and raining and thundering, and we just covered up. But we come in early and started taking chairs down in the storm cellar and told Mother, "Fix dinner early." And we all run down to the storm cellar and shut the door. We could hear tubs rattling around and big noises, but we didn't know what was happening. When it got through, we came out and went in the house. The doors had blowed open. Bessie always went to the piano and started playing first thing when we would go in the house from anywhere. She went and "Oh listen!" she said. "It just sounded so pretty." We looked, and dust had blowed out of that piano. And it was just balled-up stuff out of that piano.

It was late in the evening, so we went to bed. Of course, you went to bed at dark then. It was just raining really hard, and I heard somebody hollering, "Mr. Henry! Mr. Henry!" I called Daddy and said, "Somebody's at the front gate." "Aw, surely not," he said. And we got up and went to the door and it was our neighbors, Mr. Pierce and Uncle Grundy.

"Is everybody all right in here?" they asked. They said there had come a tornado and told all the things that had blowed away. Daddy went with them then, and they went around to check on everybody.

VIOLET HENSLEY

The Lord would give us rain. There were times that you didn't get rain and the crops were burning up, and we were worried about it. But luckily, the Lord would give us rain. We always came out with enough to get by on. We had to ration it between all the animals and ourselves. In the summer, they got grass, etc., and my dad taught us how to feed them. During the winter, the animals got pretty thin—and we got pretty thin, too! We depended on the Lord—we just prayed and hoped it rained before the corn completely died. I'm sure our faith helped get us through tough times.

MARY FRANCES LOVELL IZARD

A farmer was always at the mercy of the weather, which could have a devastating effect on our crops. Especially in hill country where we lived the soil tended to be thin with lots of rocks. Like the seed in the parable that fell among the rocks, as soon as it came up, it withered away because it lacked water. Thankfully there are some plants that will thrive even in hot dry weather. Maybe that is why the people in the South ate so many black-eyed peas, okra, and butter beans.

Our house was situated in a small valley surrounded by hills. A small creek ran behind the house. The garden soil was much richer and could stand the drought better than the land on the hillside. Also the garden was started as early as possible and there was usually quite a bit of it harvested before the summer and hot dry weather hit. If the garden yield was less, there was always a second chance with the fall garden.

Damage could also be caused when there was simply rain at the wrong time, such as when the hay had been cut but was still in the field drying, or a thunderstorm with a lot of wind could blow down corn stalks and cause some loss. However, these tragedies did not happen very often. But during a time like the Depression, there was nothing in reserve if the crop failed.

GRACE FERGUSON JOHNSON

During the Depression gardens and crops just burnt up. Me and Austin wasn't married during the worst of it. First the gardens started burning up and we just let 'em go, that's all we could do. People didn't water then. Then when our crops went to burning up, our corn and cotton just burnt up! We thought then that we'd cut the corn and shock it while the little shoots that was ears in there was still green. Then when it dried, we'd feed it to the stock. After doing that, when we went to pick the cotton you couldn't pick it. You just pulled the bolls and hauled it to town and had it ginned out. It wouldn't be a very purty bale of cotton, it didn't look like it would if it'd been ready to pick. That was about it; everybody's corn and cotton burnt up.

[Author's note: Excerpt from "Down Memory Lane." Used with permission.]

EVELYN LANGLEY

[The drought] hit in 1931 and 1932. It was as dry as a bone. With no rain for the crops, we didn't make anything. When fall come, if you had a few cotton bolls, you'd have to pull it in the burr and take it in the house. At night, you'd sit around the fireplace and pick this cotton out of them burrs. You'd put the cotton in one pile and pitch the burrs in the fireplace. And maybe you'd get a bale that way.

With no money, what did you live on? Early of a spring, to make a garden. I think we got enough rain that we made a little bit of stuff in the garden. But you know, cotton is a hot-weather plant that came on later. But it just didn't rain and we didn't make any. And what little corn we made was just a little ol' nubbin' that wasn't much good. Our livestock didn't look too good when they came through winter like them days. They just didn't have enough feed come through the winter lookin' real good.

WILLIS MAGBY

The year we came back from Arizona, 1930, was the big drought here. I don't think we hardly made a bale of cotton. Very little was made. When the drought hit, we didn't have much. We had a drought in 1929. When we picked the cotton, we could see the tracks where we plowed two or three months before. It hadn't rained enough to wash away the tracks. It rained in June and the cotton was plowed after the last good rain. So the cotton was virtually made before the drought started. And so was the corn.

We used to have gristmills around the community that would grind our corn for an eighth of the mill as a toll for grinding. We had one at Beaton, one at Marcus, one at Point Cedar. But when you had a bushel of corn and they ground it, you would still have a bushel because it puffs up. It was customary to take the corn over to the mill every Saturday and have it ground into meal. Many times in my lifetime, we didn't have any money to buy flour during the Depression.

After we planted, in 1939 the Caddo River got over the thirty acres of bottomland we rented from Charlie Wade, who lived in Arkadelphia at that time. We wound up having to replant the corn. It was late, you know. We made some corn. [chuckling] You can get by on a lot less than you've got figgered.

So what would you do? We'd have corn bread for breakfast. It wasn't our choice. Many times I've had that. I still like corn bread but I don't like it for breakfast!

MAMIE H. MAYS

The droughts were terrible. Your stuff just wouldn't grow. If it did grow, it didn't amount to much. This was around 1930 when the Depression started. All the farmers were in the same fix. They did everything they could to make a dollar or two. But we always had a good garden. Sometimes, the sharecroppers were not allowed to have a garden so Daddy would give them vegetables from our garden. They told us later that if Daddy had not given them food, they would have starved to death. Daddy was always goodhearted with everybody.

What about your cash crops like corn and cotton? We made the crops but we didn't get much for it. Five bales, for example, wasn't very much, But if you could make ten or twelve bales of cotton, you could make it through the winter good, you know. A bale would be between four hundred fifty and five hundred pounds. We'd get seven cents a pound. At one time there was a cotton gin in Lono, which at one time was a real prosperous little place. We even had a doctor at one time. Now there's only one store in town, some churches and houses.

LLOYD A. PERRY

Drought could keep a farmer from making anything. A summer rain could mean the difference between plenty and poverty. The worst drought years in this area were 1930 and 1934. There have been other droughts, one especially bad in 1952, but those Depression droughts wiped this county out! It was hard to find anything to eat, and there was nothing to feed the cattle. People went through the woods, chopping down the brush and saplings so the cattle could eat any green on the wood. The poor, starving animals would glean every bud from the twigs. Finally, some cottonseed, hulls, and meal got shipped in and them as could get some of it fed a little of that.

It took about thirteen hundred pounds of picked cotton to make a bale. During some of the bad years, a farmer might have enough cotton for a quarter or a third of a bale, so he and other farmers with small amounts of cotton would go together to make a bale then divide the proceeds—just one more way to scrap together a few dollars to help them survive.

UNION H. STOUDAMIRE

Some days were so dry, people's corn burned up. Even our garden burned up. But we would always get enough to survive on. And 1931 was the year that all the locusts invaded the country. You could hear those things at four o'clock in the morning. People didn't have feed to give to the stock. They cut the corn stalks down and everything. Gathered their own hay out in the fields, hauled it, and put it in the barn.

Maybe this man didn't raise much. Maybe that man over there raised better. He would divide with this one over here. We would always divide. If you didn't have it, and we had it, we would divide with you. We lived together. We survived! We survived! I don't know how we did it but we did. God was just with us.

ARWILDA WHITESIDE

In droughts, Father prayed for rain, then it rained. We were in Clarendon, Monroe County, out in the delta, at the time. The crops would just burn up.

When the 1937 flood came, the water was coming right up over the fields to where we lived near Big Baptize Lake where we baptized people. We were told that anybody who lived on the bottomland near the lake had better get out of town. Daddy stayed home but the rest of the family had to move from Clarendon to Brinkley by mule and wagons, some thirty miles away walking and taking turns riding in the wagon the older children. Not long after we left, the levee broke and washed away a lot of the houses but not ours. We were blessed because we lived farther away from the lake than some of our neighbors.

The wagon was loaded on with all we had. It was cold, wet. We cried, but we made it there. Papa found us somewhere to live. We stayed there two weeks. We then went back home by wagon, and walking when the water went down. We continued getting the crop out from what was left.

Chapter 21

Sources of News, Listening to Radio

CELIA ALMYRA GRAHAM ACREY

Before we got a radio, we'd go up to our neighbors and listen on Saturday nights to the Grand Ol' Opry. Pretty soon after that is when Papa got a radio. It was battery operated and we kids couldn't fool with it at all. Papa would wait until he could get through outside. When he come in, he could turn it on himself, but he didn't want us foolin' with it. He'd listen mostly to just what he wanted to hear. He got the radio just for him, I guess.

KATHRYN BAJOREK

What news? The only news I ever recall was one time my dad had a radio that operated off a battery. We were never allowed to turn the radio on. Only my daddy could do that. I remember it was during World War II and he was listening to news about a conflict or an election and my dad would just pray that night and we couldn't say a word. None of us could move or say anything because he had his ear right up to the radio because the battery was almost dead. It didn't make sense to us anyway because we didn't know what was going on.

My mother would trade two dozen eggs for the *Progressive Farmer*. My dad couldn't read so after we'd go to bed, my mother would read the magazine to my dad. She also read the Bible to him every night.

She also used those newspaper pages to make string quilts out of inch-wide scraps left over from making our clothes. She would take these strings and sew them crossways on a pedal sewing machine. After she got them all connected, she kept them straight by pressing them between four pages of the *Progressive Farmer*. After putting the strings together, you had a beautiful quilt top.

BESSIE BLACKNALL

Daddy took the weekly Kansas City Star. He loved to read and went through the whole Bible. Mama said he was going to go blind because he read so much.

Radio, favorite programs. Grand Ol' Opry, Gabriell Heater, Lum & Abner, Amos

& Andy. My uncle got a radio and everybody in the community would gather at his house. In the evenin', when we quit the field, we'd go up and listen to that program on his radio.

Did it bother you that Amos & Andy *made fun of the way black folks talked?* It was okay with me. I thought it was kind of funny.

LOUISE J. BRANT

We had a battery-charged Zenith radio. About every three weeks, we had to take it into town to have the battery charged. When that radio was gone, we really missed it.

Favorite programs. Amos & Andy, Lum 'n' Abner (we had a couple of our dogs named "Lum" and "Abner"), Bob Wills and the Texas Playboys—Oh! And Roy Rogers. If we were working in the fields, we would run into the house at noon to listen to radio. For newspapers we got the *Ozark Spectator* and the weekly *Kansas City Star.*

DOYLE L. COLLINS

Sometimes we would go to a neighbor's house—this was before we got our own radio—to listen to Joe Louis fights and political conventions. Even though I was only seven years old, I remember going to a neighbor's house in 1936 to listen to the Democrat National Convention. Senator Alvin Barkley was giving what, I think, was the keynote speech, and he was later the vice president under Harry Truman, you'll remember. After 1938 or '39, we got our first radio.

What were your favorite programs? We listened to such programs as *Fibber McGee and Molly, Grand Ol' Opry, Bob Hope, Gangbusters, The Shadow.* I remember particularly on Monday nights, we would "look" at the radio as we listened to the *Lux Radio Theater.* Cecile B. DeMille, that famous director, was the host.

Newspaper. For years, we received the weekly *Kansas City Star.*

WILLIAM EDWARD DELAMAR

We didn't know too much. No radio, no telephone, no nothing. People who were able had telephones, sometimes maybe a mile from where you lived. The old kinds with a crank on them, see . . . ? [demonstrating the crank motion] And if somebody would call you and say somebody wanted you on the telephone, you would walk that mile to the telephone. We just had conversations and enjoyed ourselves that way.

JEAN EDWARDS

What were your favorite radio programs? Country and western music ("hillbilly"), the *Portia Faces Life* soap opera, *The Green Hornet, Amos & Andy, The Shadow,* with its screeching door. Sometimes on Saturday nights we would listen to the *Grand Ol' Opry*. But we would prefer a story over *Grand Ol' Opry* music. Agricultural news, boxing (Joe Louis), and religious programs like the Stamps-Baxter Melody Boys quartet. They'd come on and sing "Give the World a Smile Each Day." When their program finished, it was time to pick up our hoes and go back to the fields to chop cotton.

ROBERT EUBANKS

Of course, these were the days before television. We had a family radio that was used on Saturday nights for the *Grand Ol' Opry*. We didn't have electricity in the home so we had this one battery. When the battery would go down, we had to take it into town to someone who had a battery charger. We'd get it charged up and then it would be good for another month.

LaVERNE WILLIAMS FEASTER

I shall never forget the newspaper. We received the *Arkansas Gazette* as far back as I can remember. It was delivered a day late on the train that dropped off the daily mail. Our dad would walk from the store to bring the mail and the *Gazette*. We were not allowed to read the funnies until we read the news. Daddy would quiz us about news articles when he came home.

VERNON L. GLENN

We didn't get the news at home. We got it—you know somebody might come down the road. My dad would go out there and talk. We really didn't get much. We didn't have a telephone or radio. Really, we didn't know what was happening. If there had been a tornado over at Walnut Ridge—thirty miles away—we would not have known. It was just word of mouth.

I'll tell you this story. There was a woman named "Serilla." We called her "Rilla." She was one of those women that had all her belongings in a sack and she carried it around. People knew her and would let her stay with them. So, she was our telephone line, because she would live at Lynn with my uncle and family, and she would get all the news and head up the road and we were glad to get her, 'cause she knew all the news. She would tell all the news, who was sick, etc.

ZELMA LOUISE HAMILTON

We got our news through the mail and word-of-mouth from neighbor to neighbor. We didn't have a radio but the neighbor next door did. We would go down at six p.m. and listen to *Amos & Andy* when our parents would let us. One evening my brother and I stayed too long. Daddy had told us to get back home by seven o'clock that evening. But the lady was making candy and my brother wanted to stay and eat it. I tried to go home but some boys threw snowballs at me to keep me from leaving.

When we got home, Daddy said he had been on the front porch calling for us, but we didn't hear him because those boys were throwing snowballs at me. He whipped us both with ropes used for plow lines. I tried to tell him why we were late but he slapped me so hard that I could not say anything. That was the one and only time my daddy ever whipped or touched me in a disciplinary manner. After this, when I needed to be disciplined, he told my mother to handle the punishment.

MARY FRANCES LOVELL IZARD

In a small close-knit community where most people were related by blood or marriage, news traveled by mouth. We also got news either at church, at the local store or even funerals, any place where people got together. The neighbors came by and shared news with us, but mostly we were so isolated that usually we heard only the most important events. There was no newspaper delivery. Occasionally, someone came by selling the *Grit*, a weekly newspaper. My dad always bought one and read it from cover to cover. Since we believed in recycling, it ended up along with the out-of-date Sears & Roebuck catalog in the outhouse.

We had a battery-powered radio, but most of the time the battery was run down. If we could get it to play we listened to the *Grand Ol' Opry* and a station in Del Rio, Texas. Batteries were not easy to replace due to shortage of money. Therefore, the radio was not turned on except at certain times of the day when the news was on.

WILLIS MAGBY

We got a radio in 1941 I believe it was. Very few people had radios around here. And naturally, when we got a radio—which was when I was almost twenty-one years old—was a big deal. Saturday night was the *Grand Ol' Opry*. It was good entertainment, better than most of what we get now. When I lis–

STORIES OF SURVIVAL

tened to the radio, I could visualize people. Even some of the daytime soap operas, my mother would listen to them and if I was not working or doing something, I would listen in the afternoons and it was almost like watching a movie. And we "watched" the radio! I listened to *Amos & Andy, Lum 'n' Abner, I Love a Mystery*, and the squeaking door on *The Shadow*. There were no bad ones.

DULAS L. MASSEY

We didn't even think about getting news. One of my dad's aunts took the Sunday *Arkansas Gazette*. All I knew was what I read in the funny papers (comics). My granddaddy had the first radio and he put it on a table and pointed it out towards the window. He's keep the window open with curtains on it. He would turn it on and in the summertime the yard would be full of people listening to the radio because he kept the volume up so everybody could hear it. *Amos & Andy*, the Jack Dempsey and the Gene Tunney fight. The yard was full of people. The *Grand Ol' Opry*, that's what everybody liked.

MAMIE H. MAYS

Once a week, we got the Kansas City Star *by mail*. Me and my brother would read that all over. We couldn't believe some of the things that were going on in the world.

We didn't have a radio for a long time. My aunt Ida Mae Heard was the first to get one. She was postmaster in Lono. Every Saturday night, she'd let all the young people come there to sit real quiet and listen to the *Grand Ol' Opry*. The house would be jammed full. It would be so quiet and nice. That was our Saturday night outing because there wasn't much to do in the country. Aunt Ida had polio when she was young but since nobody knew about that at that time, they just called it the "fever." It left her with one of her legs not growing any more. It also affected her mind.

EVELYN M. COONFIELD METCALF

Newspapers were used for many things. If you could of found a newspaper, it was read and reread. We used newspapers and straw under the carpet in our front room to help to keep the cold air out. It was glued inside the frame of the inside walls and covered with a small board to keep the cold from coming into the house.

Radio was great! The radio was a special item in anybody's home. I was going to Eagle Corner School when my daddy finally ordered our first radio from Sears in 1941. I watched the mailbox every day for the package. My dog Pooch would lay down and let me stand on him to reach the mail. The day the radio came I heard a car horn. Pooch and I ran to the mailbox, and there was our radio!! The box was too big, though, for me to carry, so I told Pooch to go to the house and get my mother. She came out and picked up the box, and I carried the mail to the house.

Favorite programs. Of course, we had our favorite programs. *Fibber McGee and Molly, The Green Hornet,* soap operas (*Ma Perkins*), *The Lone Ranger, Amos & Andy.* The *Grand Ol' Opry* was on Saturday night and sometimes our neighbors who didn't have a radio would come to visit and listen to the *Opry.* We would have popcorn, tea, coffee, maybe Kool-Aid for us kids. When the news came on, though, it was my dad's time. We didn't talk or make a noise. We knew to be quiet "or else."

WILLIE MORRIS

We didn't have no electricity out in the country so we used a battery-operated radio. The battery would be about four inches wide and a foot long. It would last for six or eight months, something like that, before you would have to buy another one. Our favorite programs were *Amos & Andy* and the *Grand Ol' Opry.* The women liked *Stella Dallas.*

Were you bothered by the way African Americans were portrayed on Amos & Andy? Nothing against the radio programs. They were good. Back in that time, the way blacks were portrayed was fun. After we got a little older, when it got on television, we seen what kind of show it really was.

[Author's note: Mr. Morris is African American.]

LLOYD A. PERRY

We began to see an occasional radio in this country beginning about 1937–38. Kids met on Saturday nights where there was one and visited and listened to *Lum 'n' Abner.* A couple of us rode five miles on a mule to hear the fight between Joe Louis and Max Schmeling. I hate the new so-called country music. It used to tell a story, made some sense.

WILLIAM PIERCY

We received the weekly Kansas City Star *every Wednesday, and the* Benton County Democrat *every Friday.* During WWII, we got our first radio. It was battery operated. Our favorite programs were *Jack Armstrong* ("The All American Boy") and *Fibber Magee and Molly,* the *Grand Ol' Opry, Louisiana Hayride, The Shadow Knows* (with its creaking door that would send chills down my spine). We'd listen to Gabriel Heater to hear about the war.

Crank telephones were another way to know what was going on. We had a community telephone system using crank telephones. Each family had its own rings—for example, two long rings and one short ring. All the telephones rang when someone made a call but you would have to identify your own ring by the length and number of rings. It worked!

UNION H. STOUDAMIRE

We had one postman who brought the newspaper in a horse and buggy. He had a whistle he would blow when he got to the mailbox. We lived back off the pike, but when we heard that whistle we made it to the box to get our mail. We didn't never get nothing but the newspaper, though.

JAMES A. THOMPSON

My granddad sold maybe a hog or a cow. I don't know where he got it—maybe at Harrison—but he got an old battery-powered radio. We didn't have electricity. The boys had a big old walnut tree in his yard and they climbed that tree and put an antenna at the top. It worked fine but he would turn on that radio only on Saturday night. That was the only time he would use it so he could listen to the *Grand Ol' Opry*. The neighbors all found out about that so every Saturday night, here they would come. He would have ten or fifteen of his neighbors who would come from quite a distance. My granddad would let the radio play for an hour and then he would say, "Folks, the show's over!" and would turn it off until the next Saturday night.

EVA SMOKE WELLS

News about the Depression was coming to us through the grapevine, maybe two or three months after it had happened. My dad subscribed to the weekly *Kansas City Star*. What we were most interested in was what quilt pattern was always in it. And the feed companies started selling feed in the most beautifully colored sacks. When Daddy would bring home feed sacks, it was always an argument about who was going to get them. In my quilting magazines there are places advertised that make reproductions of the patterns used in the feed sacks. If you are fortunate enough to own an original, you are envied.

Favorite radio programs. Daddy would rather listen to *Inner Sanctum* and *Fibber McGee and Molly*. I liked the morning shows—*The Breakfast Club With Don McNeill*. And of course I liked to listen to the "soaps." Stella Dallas, etc. If the battery would run down, we would all chip in to get a new one. I think the battery cost nine dollars.

LUCILLE RIDER WILSON

My grandparents subscribed to the Progressive Farmer, *the weekly* Kansas City Star, Farm & Ranch, *and we bought* Grit. I read them from cover to cover. That was what I read except for what I got at school. If there was a public library, we didn't know about it. The *Progressive Farmer* and *Farm & Ranch* had "housewife" sections.

How did your reading benefit you in later years? I think it helped me see the world a little bit bigger. To read about people who were different from me, lived a different lifestyle. It broadened my world, for sure. My family did not show emotions. I don't remember ever being hugged or told that I was loved, anybody holding my hand, or anything like that. So reading opened up the world to me.

Listening to the radio. After we came in from working in the fields, we would gather together and would listen to the radio. *Jack Armstrong,* soap operas like *Ma Perkins, Stella Dallas.* Bob Wells played on KVOO out of Tulsa, Oklahoma. We would come in from work at noon and he had a thirty-minute radio program. We listened to him while we were eating lunch. Even before we had electricity, we had a battery-operated radio but you didn't listen a lot because you would run down the battery. When the community first got electricity, we didn't because Daddy didn't think we could afford it. When we did get it two or three years later, the first bill was two dollars and fifty-three cents. [laughing]

Chapter 22

Christmas Memories

DOROTHY COX COSTON ASHLEY

Cakes, pies, fresh ham, and fruit. Mother always baked two or more cakes. Her Lady Baltimore cake was my favorite. She always boiled a whole fresh ham which lasted several days. We produced our own pork and killed our own hogs each fall. As gifts, we always received a bag of Red Delicious apples, a bag of oranges, a bag of tangerines, and a couple of whole coconuts. Before the coconut was broken open and the fresh coconut extracted, we drained the coconut milk and drank it. We were allowed to eat the fresh coconut as a snack, but Mother grated some of it and used it to make pies and cakes.

Making "monkey heads" from coconuts. After the coconut was extracted from the shell, we glued a piece of red fabric inside the half of the shell that had the "eyes," soft spots on the shell that resembled a mouth and two eyes. A lighted candle was placed inside the shell and we called it a monkey's head. I never really understood much about the supposed monkey's head, but Daddy enjoyed doing this at Christmas.

KATHRYN BAJOREK

I never got any Christmas except a brown paper sack from church with nuts, fruit, and candy. Never had any birthday celebration except my dad thanking God for our blessings. I never got any Christmas except the church gave me a sack. In that sack was what we then called a "nigger toe" [*Perjorative slang for brazil nuts, which were said to resemble a black man's toes*], an orange, an apple, and a piece of peppermint candy. I always looked forward to Christmas because I knew the church was going to bring us a sack! There are so many things that our children are missing now.

BESSIE BLACKNALL

When we were young, the night before Christmas we had a hard time getting any sleep. We were waiting for Santa Claus but sometimes we didn't get anything. We were told that Santa had hit a mud hole and had to go back.

Do you remember a time when as a child you did get presents for Christmas? Oh yes. I got a little doll. The boys would get little pistols. Sometimes we would get pine straw, stick it into water, and put flour on it and hang it out like mistletoe on the front porch. We'd get an apple or an orange or one stick of peppermint. Then we would try to out-save everybody else to make them ask for it.

FRED NORMAN BLANKENSHIP

Our parents would get kids off to bed early on Christmas Eve so Santa Claus could come. During the night Mother would hang up all our stockings. We'd get up on Christmas morning. We would get an apple or an orange or pecans, English walnuts, and stick candy. I remember the first present I got besides something to eat was a little one-bladed knife with a chain on it. I wish I had kept it. I never enjoyed Christmas as much after I found out who Santa Claus really was.

R. L. "BILL" CARTER

I didn't get a Christmas present until I was a teenager—a set of dominoes. We played those things until we wore the dots off. At Christmastime, we usually got an apple and an orange and three pieces of candy. And that was it. We didn't get nothin' like that in between. You could buy a big yellow bar of "O.K." soap for a nickel.

What does a bar of soap have to do with Christmas? When you're poor folks, this has everything to do with Christmas! We'd draw names at the old schoolhouse and most of the time we would get a ten-cent handkerchief for a Christmas present.

MARY LEONA CARTILLAR

We didn't get anything for Christmas. We always had fruit salad on Christmas. Nothing. If we cut down a Christmas tree, the decorations would be something we made, you know. We didn't always have a Christmas tree, but Mama always managed to get the fruit for the fruit salad. But we didn't feel deprived. No, it was just a way of life, as far as we were concerned. I guess that's something that stuck with me. We always knew we had to survive one way or another.

GENEVA LOUISE BECKETT COTTON COLLINS

Christmas candy, nuts. Just before Christmas, my dad went to Paris, Arkansas, where he bought everyone a gift plus Christmas candy, nuts. My granny would take coconut hulls and fill in the holes to make eyes, a nose, and a mouth. Then she would add scraps or yarn for hair. My brothers would take

78 rpm records, heat them, flute them, decorate them for catch-all bowls. Dad was wonderful at carving figurines—cats, dogs, horses, and puzzles.

Bedding down overnight company. We always had a crowd at our house at Christmas and other holidays like the Fourth of July and Thanksgiving. Some were overnighters. Pallets were wall to wall. It never was hard on Mama. There were always willing hands and feet. Daddy bought a huge candy cane at Christmas, which he would break up into small pieces. In the house everything was coming together, no one came empty handed. They brought different meats, canned goods, and lots of baked goodies.

WILLIAM EDWARD DELAMAR

People back then thought there was a Santa Claus. Children don't believe that anymore. We would hang our stockings around the fireplace somewhere and the kids would go to sleep. Then our parents would put apples, oranges, Brazil nuts, or whatever toy we might get, things like that, into the socks. When we would get up in the mornings, getting what was in those socks would be the first thing we would do, and we were just as happy as we could be. And we would wonder how Santa Claus got into the house. They would tell us that he had come down through the chimney and we didn't know why he didn't get burnt in the fire.

LEONA STITH DUFFY

It was getting close to Christmas, and we had about three cups of flour in the house, no meat or vegetables. All we had left was some corn. We had three children looking for Santa and we had no money. As the time got closer, we silently prayed and kept up our spirits before the children. Finally, Christmas came.

[My husband] O. C. and I went in the other room to discuss the matter. I said what will we do? It seemed if we were beaten this time. O. C. said he would swallow his pride and borrow $2.50 from our neighbor, C. C. Jackson, to get the children some toys. I said I thought we had enough flour and sugar to make a plain cake and we could shell a bushel of corn and O. C. could take that to the mill. O. C. added that we would also kill our pig. He was poor but it would give us some meat. So we would have meat, bread and cake just the same. We already had a cow and she was giving milk.

"After our plans were set, O. C. took the corn to the grist mill and had it ground into meal. He brought it home and that gave us meal for our bread. In the meantime, there was as family of white people, the Pages, that lived about two blocks away.

"The older lady became ill and asked me if I would help them with their baking. She said she did not have any money but she would give me a chicken . . . . I had baked the plain cake before I left, and it turned out fine. O. C.'s aunt and her friend came to see us that evening and ate just about all the cake, not knowing that was all we had, but another miracle happened. The lady I worked for liked my work so well that she gave me a big plate of cake with slices of all kinds she had baked and also a live hen. So we had cake for Christmas just the same.

"I thought so much of my hen that I would not kill her. We named her 'Mrs. Page' and she learned to come to us when we called her. When I got home from work, O. C. had been to town and had borrowed money from Mr. Jackson and bought the children some toys: Little 10¢ horns, little 25¢ dolls, some apples, oranges, and candy. So we had a good Christmas after all."

[Author's note: Excerpt from "The Eternal Dream" by Leona Stith Duffy. Used by permission.]

GENEVA KING EMERSON

Cooking Christmas dinner for the family. Whenever all the aunts, uncles, and cousins gathered at my great-grandparents, Grandma cooked whole hams plus many pies or cakes baked before the big day. Vegetables and fruit cobblers simmered in large pots, even dishpans, on the woodstove. Her oven was immense with double racks from which she took great pans of breads and other baked foods like sweet potatoes. Their family was grown and she and Grandpa were good, experienced providers. During the lean times, that was their way of helping their young families.

Seating arrangements. There wasn't room at the table for a fraction of the group. The elders and mothers with very young little ones to help feed got the seats of honor there. The rest of us found a place, sitting on the steps or in groups under trees. I liked to sit with the big kids and young couples on the edge of the porch with my feet dangling off it. I don't know how Grandma found enough plates, saucers, pie pans, etc., for us to eat from. There were no paper plates and plastic ware back then.

Mom encouraged our creativity with holiday preparations. At Christmas, we searched for a beautifully shaped cedar in our fields at home, then decorated it with strings of popcorn and winterberries and cut-outs from any scraps of pretty paper we were lucky enough to obtain. Mom would bake cookies and we might hang some with a string on the tree. If possible, she would make us each a doll or toy animal from flour sacks and old clothes and stuff it with cotton from our fields.

ROBERT EUBANKS

One of my earliest memories of Depression times was at Christmastime. My father had one five-dollar bill to buy Christmas for the whole family. He played a trick on us. He put that five-dollar bill in his book of cigarette papers and hid it. We never did find it. We looked all over the house for that five-dollar bill. When Christmas morning came, we weren't expecting anything, but when we got up, there was this Christmas tree decorated and there were presents under the tree for everybody. He bought Christmas for a family of seven kids for five dollars.

I had a metal toy airplane that was very substantial. I kept it for years and years. I don't remember what the other kids got but everybody had something. In addition to that, we had Washington apples, bananas, coconut, everything needed to make a good fruit salad. Of course, we had a ham out in the smokehouse so having a Christmas ham was no problem. We had that great fruit salad, and Mother would cook. She would start making candy about Thanksgiving time. She had a big ol' five-gallon wooden box that I think contained Rex jelly, kind of an artificial fruit jelly. It was inexpensive and you could buy what looked like a five-gallon bucket of it. She would use the box to hold the candy.

Mother would make divinity, taffey, toffey, chocolate fudge, some with nuts, some not. The nuts would largely be peanuts or hickory nuts. Pecans were not in that plentiful supply. We'd get a few English walnuts to go in the fruit salad. I remember sneaking into the pantry where the food and the candy box [were to] get me some divinity, which was my favorite. I'm sure she knew I was doing it, but she never let on like she knew it.

MENARD EULIS HARVELL

How we got our Christmas money. Right after World War II it was a rough time [in Izard County]. We built a house over yonder the year after the war closed. Well, it come Christmastime and we didn't have no money; didn't have NO money. My daddy said, "We ain't gonna have no money for Christmas." He said, "Let's go down the road from here and cut some logs, and see if Bill (Warnett) will haul 'em." We went down there and dad and I cut the logs and Delbert took an ol' mare, and he'd skid 'em out. And we got Bill to haul 'em. Out of two loads, our part was $25, and I guess Bill's was about like that. And that's the way we got our Christmas money. We bought apples, oranges, candy and coconut. That was about all our Christmas.

It wasn't Christmas without coconuts. You'd punch a hole in that thing and drink that water out and wasn't that good! They'd pour it in a glass and pass it around and let everbody get a sip. Sometimes you'd take a little bit bigger sip than you was supposed to. We made dippers out of the coconut shells and put handles on them and used them for two or three years in that water bucket.

[Author's note: Excerpt from "Down Memory Lane." Used with permission.]

GEORGIA HEARN HILL

We didn't celebrate birthdays, but oh, my Lord! Christmas was one of our greatest days! Mama would start baking cakes about a week before Christmas. We already had the chickens and turkeys. After chopping off their heads, she would put them in hot water to pick off the feathers. The chickens would be so fat and would make the best dressing! After she got the cakes all baked, she would wrap five or six cakes—coconut, chocolate, all kinds—in white cloth and put them down in a flour barrel to keep them moist.

We would cut down a Christmas tree, make decorations out of crepe paper. And we made popcorn balls and peanut brittle out of molasses. One Christmas Eve night, I woke up, Mom and Dad were still sittin' up. I looked up and yelled, "There's old Mary (my black doll) in my gift sock!" Daddy said, "Santa Claus just left!" Then he slipped outside and threw a stick of wood up on top of the house. I was so scared I could hardly breathe because I thought Santa Claus was about to come back in the house and take away my doll!

Anyway, I got up the next morning and my doll Mary was still in my stocking. I'd played with her so much that she was all torn up! That's what I told Mama and Daddy I wanted for Christmas—another Mary. And that's what I got. The stockings would be full of candy, apples, oranges, and nuts. My brother would have BB guns, marbles, Roy Rogers stuff in his stocking.

MARY FRANCES LOVELL IZARD

Christmas was a very special day. Not because we received a lot of gifts and toys because we didn't. We always went with my dad out to the woods to find a cedar tree to trim for Christmas. The hunt for just the right tree was as special as any other part of Christmas. When we found the right one and cut it down, Daddy put a little stand on it made from a couple of boards. Then we trimmed it. Then it turned over and we had to tie it to nails in the wall! We had a few glass balls and some tinfoil icicles and some old garland that was slightly mashed from being stored so many times, but we didn't notice that at all.

The day before Christmas Daddy would go to town. When he came back he had a big paper bag. That night we all set a shoe out by the fireplace and the next morning there was an orange in our shoe and some hard candy. That doesn't sound like much today when we have all sorts of exotic fruit all year round, but an orange was a very special treat that we seldom ever got. If there was any extra money the children would all receive a small gift such as gloves or a toy.

RAYBURN THOMAS JOHNSON

I growed up in the Depression, and there was very little that you got for Christmas. We'd get apples and oranges and a little candy and that was about it. I had one grandma till I was about six or seven years old, then after that my mother and dad died before I married. But my aunt Omie Hamm made boxes—so big [demonstrating] of the best homemade candy you ever ate in your life! And she'd send us a pretty good size little box of that every Christmas. She lived at Bear Creek, and maybe she'd give us a handkerchief, or something like that and boy, we thought we was really getting something! I remember one time she give me a little ol' ball and now, boy! We thought that was something!!

[Author's note: Excerpt from "Down Memory Lane." Used with permission.]

EVELYN LANGLEY

Christmas tree at different homes. Oh, Christmas was the time of our life. We had a Christmas tree at different houses. One family this Christmas, another the next. This Christmas tree would touch the ceiling. It would be so big that it would bow over in most houses. We'd all take our gifts. We'd make our decorations. We'd cut out paper, make chains, and glue it together and put the chains all around the tree. We'd pop popcorn ropes and decorate that tree.

We didn't have no electric lights or things like that on it, but we didn't know the difference. We didn't know anything about having electric lights on it.

We'd all meet at that person's house Christmas and take food, kinda like we call potluck nowadays, you know. We'd eat, have a Santa Claus. Some man in the county would play Santa Claus. My daddy was a whiz at it. [laughing] He'd call the names on the packages. They wouldn't be wrapped in pretty Christmas paper like we'd have nowadays. It might just be in a paper sack.

There were some men in the community that could pick guitars and [play] a fiddle. They'd play Christmas carols.

Christmas gifts. The parents would bring their children's gifts. Us kids didn't give gifts to our parents because we didn't have any money to get 'em with. Maybe they'd give us some socks. Mama would knit us some gloves, stuff like that. And that would be a nice gift to have gloves to wear to school, we had to walk so far in the cold. When we'd get to school, lots of times our hands would be so cold our teacher wouldn't let us get real close to the stove because if you did and got them cold hands hot real fast they'd hurt real bad. So we had to just stand back and heat 'em slowly.

WILLIS MAGBY

Gifts we received. I always got a pocketknife. I'd lose it usually in about a week. Sometimes it would take longer but I would never have it by the next Christmas. We'd always get some fruit, and firecrackers were a must. We're talking about eighty-five years ago. Beaton and all the other stores around would stock up on fireworks on Christmas. We wanted to shoot them immediately. We'd have just as good a Christmas then when I was small, as we do now.

How do you account for that? Some people are going to be happy with whatever they get. I've lost lots of family, you know. My younger brother was killed when he was twenty in New Guinea during World War II. On Christmas Day in 1968, my younger brother died at our domino table with a heart attack. But I have had so much. I've had a rough time through life, but I've had lots of joy to make up for it.

EVELYN M. COONFIELD METCALF

Christmas was always celebrated on a Sunday at my grandparents' large house. I had lots of aunts, uncles, and cousins, and each family brought their favorite food and their own plates and silverware. This was before paper plates and plastic silverware. One of my great-aunts made around three dozen rolls and baked them at my Grandmother Coonfield's house, and we had homemade butter for the rolls, along with honey from one of Grandfather's five or six hives.

Christmas menus. We usually had chicken, turkey and all the trimmings, sweet potatoes, white potatoes, peas, green beans, corn, and beets. The beets and turnips were kept fresh in a hole dug in the ground and covered up with dirt and straw. All the food came from our garden and orchards. Then were the wonderful desserts: pies, cakes, cookies of all kinds, candy and FUDGE!

Transportation provided for families. Some of the families didn't have a car or horse and carriage or any way to get to my grandparents' house, so Grandfather loaded the wagon down with straw and blankets and went to pick them up. It was always fun riding in the wagon. Grandmother had hot chocolate and one cookie for each of us on the wagon. SO good!

Entertaining the kids. We always had music and there were cards and other games to play. There was usually an apple or orange for each of the children—and sometimes BOTH! We all enjoyed getting together with each other. In those days, children were supposed to be seen and not heard, so we went off to play in the bedrooms. The boys had one room and their fun time, and the girls were in another. The babies were usually in our bedroom for us to watch over. I was the oldest, so that was mainly my job. We were poor, but we didn't know it, because we were SO rich in other ways.

WILLIE MORRIS

Back in those days, do you know how many toys we would receive for Christmas back in my day? None. Homemade ice cream, potluck get-togethers; friends and family dinners. We made our own toys. We'd take a log that had been cut down, cut it into strips that were about two inches thick across the top and maybe twelve or fourteen inches high. We'd then bore holes through the center of them to make our own axles and stick them through the holes to make wooden wheels for our own wagons. We'd also make wheelbarrows out of 'em. Since we didn't have no cars back in them days, we would take the wheelbarrows to the plantation's commissary store and bring groceries back in 'em.

How we got our Christmas trees. We went into the woods and cut down pine trees, cedars, or holly trees for Christmas. Most of the time we got toys, BB guns, Jew's harps.

UNION H. STOUDAMIRE

The children would get together and we would go from house to house, to my kinfolks and cousins on Christmas and get fruit like oranges and apples from the neighbors. But mostly we were just having fun. We weren't looking for no food or nothing. I wouldn't have eaten an orange for nothin' because I didn't like oranges. If someone would give me an orange, I would swap it for an apple.

JAMES A. THOMPSON

As a child, Christmas for me was a big thing. But for my mom and dad, it was not a big thing. [chuckling] But I always looked forward to it every year. Before

Christmas, I would go out and pick the biggest cedar tree you've ever seen. My dad, he would go out and cut it and drag it to the house. Then he would have to cut it half off because it was too big. I would get very few of anything so far as presents were concerned. I'd get an apple, an orange, three or four pieces of hard candy. I would also get a little metal toy of some kind from Japan. That was about it for us at Christmas. But I always looked forward to it and enjoyed it.

The big cave at Christmas. Right there on Cave Creek at my granddad's farm, there was a big cave right on the creek, which was right against the farm. It was a big, big cave with a big opening in front. At Christmastime, the neighbors—there wasn't many of them—would go over there in that cave and have a little Christmas program. It wasn't much because we didn't have much. But I remember they would sit and sing some songs there in the cave. One neighbor would have a few apples. Another would get a few oranges. Seems like my dad always had a little candy or bubble gum or both and they would have a Christmas program there in that cave because there wasn't a house in the whole country there that could hold over four or five people.

ARWILDA WHITESIDE

We baked at least fifteen cakes for houses in our neighborhood. We went from farm to farm tasting and testing each other's cooking and sharing food. We made hog-head cheese, and eggnog with tartar, mincemeat pies, candy, jelly cakes with Rex Jelly. We used the Rex Jelly buckets to take our lunch to school.

Christmas gifts and shoe boxes. We got school clothes for the winter. We each had Christmas shoe boxes with our names on them. We would fill the boxes with raisins, an apple, and some chocolate Kisses. We each got one toy. My brothers would get Jew's harps; the girls would get little harmonicas. Mama made rag dolls and corn-shuck dolls for us, too. Our brothers would get them sometimes and tear them up just to be mean! On the first day of Christmas!

LUCILLE RIDER WILSON

We would go out in the woods and cut a Christmas tree off the farm. It would be about two feet tall. Then we would take paper, whatever we had, to make circles that we would hang on the tree. We'd pop popcorn and string it up. We would all get a present. I remember one time getting a little string of pearls. which I thought was wonderful. Mama would bake a lot of desserts and they were on the table. It seems to me they were on the table throughout

Christmas week. If we wanted to, we could have pie for breakfast because it was setting there.

Gifts. We would always get just one thing. I remember one time I got a little doll. I remember walking outside and standing by the chimney but I did not know how to play with that doll. [Her voice seemed to catch slightly as she said this.] I remember another year when I got a string of pearls. We went to visit some of Mama's family. These people were coal miners, which was short-season work—a little bit in the fall, and then they were unemployed. Then they would borrow money from Daddy. But they had a huge Christmas. They had food, they had toys, they had books, everything. But our Christmas was very sparse.

Mama would cook special meals for Christmas. One of the special things she did that I don't think anyone else did was a burnt cream pie. She would put sugar into an iron skillet and she'd cook it slowly until it was syrup. Then she would make a butterscotch cream pie filling out of that. And we called it burnt cream pie. Even after I was married, she would make that pie for me everytime I came home.

FLOYE WINGFIELD

We hung our stockings for each person on nails in the mantel. The living room had heat in it. You didn't have room for a Christmas tree. These were long stockings that you could put a lot of things in. The things we would always have would be an apple, an orange, some English walnuts, and some peppermint candy. And sometimes if you were really lucky, you might get a stick of chewing gum. That was wonderful!

One of the presents I got that was so precious—and still is at ninety-two—was the year I got a sure-'nuff pencil. The only thing I had before were these cedar penny pencils. The first time you erased something, the rubber would pop off. But I got a pencil with the eraser put on with metal. I could just erase and erase all I wanted to.

We would make Christmas ornaments and decorate the mantel board. We'd save sweet gum balls and save every scrap of foil paper we could find like off a gum wrapper. We would put it around the gum balls and make this beautiful, shiny star. Why, Times Square in New York didn't have anything on us.

JACK WOOD

Christmas was the time of the year that I really enjoyed. It was the time my mother and I were together. About ten days before Christmas, she would say,

"Jack, get the axe and we will go get our Christmas tree." Leaving our little ones with the oldest one in charge, we would take off walking up behind where we lived and we would search for a cedar or pine in the size and shape we wanted. When we found one, we would cut it down and drag it to the house.

Then Mother and my two sisters would start decorating it the next day. They would make ribbons for decorating. We had no lights and we had no presents under the tree. We did, however, have seven stockings hanging on the mantel above the fireplace. Then on Christmas morning, we would get up early and go check our stockings and in each one would be an apple, orange, Brazil nuts, and a red and white stick of candy. We actually knew what we would get but we were still thrilled.

I never received a toy until I was about twelve years old and it was a red coaster wagon. I think it cost about seven dollars. It had stakes all around. Since I was the oldest, I had to supervise it and take the little ones for a ride around the yard. It's a little different today because I know some families here in town that don't have room in their garage to keep their car because of all the toys stored inside the garage!

[Author's note: From interview conducted by Maggie King, Jack Wood's granddaughter.]

Chapter 23

Spiritual Life and Good Neighbors

DOROTHY COX COSTON ASHLEY

My parents were both very well informed of the contents of the Bible and they lived by Christian principles. My grandmother would not let my mother punish me for eating all of the English peas raw when I was sent to the garden to gather them for cooking. She said, "The Bible says you cannot muzzle an oxen while it is treading out the grain." Mother just wouldn't let me go near the garden any more when there were fresh English peas there. I loved raw English peas and still do.

Family and neighbors were really very important to us. We shared any extra vegetables, milk, butter, and eggs we had with friends, neighbors, and relatives who might not have any at the time, and they shared with us. In times of trouble, neighbors always helped neighbors. Neighbors knew they could depend on each other for anything and this was really a great comfort.

KATHRYN BAJOREK

I grew up in the Assembly of God (Pentecostal) Church. Girls didn't wear pants; we didn't cut our hair; we didn't wear short sleeves. That's just the way it was. My parents were Indian. If our hair got real long and Mother couldn't handle it, she would "singe" it. She would set us between her knees and put some fire to the ends of our hair while we were leaning over. She would put her hand on top of our head and when the hair got as short as she wanted it, she pulled her hand down and put the fire out. Mother said that was the "Indian" way. She said it made our hair healthier and shorter.

BESSIE BLACKNALL

I go to sleep praying every night. My daddy was one of the outstanding deacons of the church. Everybody gave him a good name and he tried to treat people just right. They didn't have baptismal pools back then. On the day they were going to baptize someone, that was a good day. Everybody walked about three miles to Low Freight Creek on Highway 7 (near Arkadelphia). The deacons and the persons being baptized would start at the shallow part and walk to the deeper part and dip them.

I hear the people talking about walking to the church at night. When they did revivals, you could hear them singing on the way to church. The people would get so anxious when they heard them singing, that they would pick up speed and start running to get there. This is when some of the younger people courted sometimes, with the parents walking along behind them as the boy walked the girl home.

R. L. "BILL" CARTER

My grandfather was a preacher. My dad was a song leader. They used to have Monroe County Singing Conventions in different country churches. There weren't no pianos in either one of the churches in our community. Dad had an old tuning fork. My brother's got it now. He would just hit the right pitch on it and they'd sing. The Rich choir was the best choir in the singing convention.

Did you go to singing schools? Yeah. I went to one. All the Rich Church choir members knew how to read them shaped notes. Dad would take a tuning fork to get the pitch and they'd sing those notes over two or three times. And then each one would sing their part. I used to like to listen to them singing those notes. I never did sing in the choir or anything like that but I liked to listen to 'em.

Neighbors would do anything in the world they could for you. I remember me and my dad had the mumps and it went down on us and we was laid up in the bed. A guy pulled up there with a wagon with fir sideboards on it. He didn't say hello or nothing. He just started to unload wood. People would help you back then.

LEONA STITH DUFFY

"*All of the children put a high value on belief in God,*" said Joseph Blank in "An American Family," "for they could see in their parents that it was an inner resource of strength. While on the farm, the [Duffy] family trekked four miles to church each Sunday. In very bad weather, [Mrs. Duffy's husband] O. C. held Sunday school and services at home. Strange as it seems I never went to a neighbor's house and asked for assistance and no one really knew the condition we were in. I really believed in the hymn I used to hear my father sing, "O for a faith that will not shrink though pressed by every foe, that will not tremble on the brick of any earthly woe."

[Author's note: Excerpt from "An American Family," *Reader's Digest*, July 7, 1977, 109.]

GENEVA KING EMERSON

Church and family were everything! Neighbors were about all kinfolks. They tried to help each other as much as possible. Families looked forward to fifth Sunday meetings when all the churches in the local association gathered with all-day singings and dinner "on the ground" and to revival meetings in the fall after the crops were laid by. A lot of good matches among the youth of marrying age were made then, and if success resulted from the revival, the saved often joined the church and were baptized in the ol' swimmin' hole on the creek.

GEORGIA HEARN HILL

Mama would get us ready for church on Friday or Saturday. We'd get the yard cleaned up for Sunday, Mama would cook, then everybody would get their baths on Saturday night. On Sunday morning, Mom would get up and fix breakfast and everybody would be off to Sunday School. After I got big enough, I would write the minutes as the secretary. My friend, Jessie Maye, loved to write the minutes, too, but whoever got there first would be the secretary. And I always wanted to be there first. That was a joy.

What social role did church play for you? Most times on Friday night and Saturday nights the church would have a fish fry and serve homemade ice cream. Sometimes they would catch fish; other times we would eat "whitey" (whiteing) fish. I think they would give you two pieces between slices of Wonder bread and that was the best!

How long would a Sunday morning church service last? Oh, my Lord! They would have Sunday School. Then we would have our singing service. At the singings, my three aunts would have on these big hats and they would sing and pat their feet and click their high heels on the floor. Then the preacher would preach one and a half hours! Then they would do an altar call—inviting people to be saved—and make announcements. After that, we would go home and eat dinner. Then we would go visiting our cousins, come back for A.C.E. Bible Study and the young people's meeting at six o'clock. Unless we had a guest speaker, the evening service wouldn't last very long. I enjoyed it.

What about all-day meetings and dinner on the ground? Yes. We had cemetery decorating, too. They would come once a year to the Richmond Hill Cemetery with their shovels, posthole diggers, and hoes. This was a community event. And the cemetery was so pretty. It's still there. It's where all the people are buried on my side of the family. We would also have box suppers at church.

MARY FRANCES LOVELL IZARD

Our spiritual beliefs were what sustained us during the Depression. My dad and the children regularly attended church. My mother did not go during these years because it was a long walk through the hills and she was not physically able to make the walk. She remained at home and prepared our dinner. When you have no control over the future and sometimes where your next meal will come from, God is the only one to turn to for help.

The pastor, who probably drove a long distance to get there, went home with a church member for dinner (not "lunch" back then). We were always glad when the preacher came home with us because Mama would kill a frying-sized chicken and we would have fried chicken for dinner.

However, it was not quite as good as it sounds. In those days the adults ate first and the children ate last! I know preachers can eat more than anyone else because when it came our time to eat there wasn't anything left but a wing, a neck, or a back. It is inconceivable in today's world to make the children wait until all the adults are finished.

It was customary in country churches during the Depression to have what was known as "Homecoming Day." Every church would have their homecoming on the same Sunday every year. At Kentucky Church our Homecoming Day was the fourth Sunday in August. By having it at the same time every year those who had moved away could make plans to come back and see their old friends and relatives. The men nailed temporary tables between the trees and the ladies cooked. When the day finally came the tremendous array of food was almost more than we could bear. I remember I always made for my cousin's table because she brought banana pudding.

The preacher preached all morning and we all ate too much. There was gospel singing and lots of visiting all afternoon. It was the one day we all looked forward to all year. Relatives and friends came back that we had not seen in years. It was truly a joyful time. Homecoming Day was followed by a week of revival meeting. Sometimes it was held in a brush arbor to take advantage of the cooler night air.

During the Depression the viewing of the dead family member was held at home. If the family could afford to have the body embalmed, the body was brought back to the home until the funeral. People really did "sit up with the dead." Many times my dad would go to the home of the one who had lost the loved one because someone had to sit all night with the corpse to be sure nothing happened to it.

What could have happened to it? There was a superstition that cats would harm the corpse if they could get to it. The funeral was in the afternoon at the church. However, the grave could not be dug until the morning before the funeral. It wasn't proper to have an open grave at night. The morning before the funeral all the men of the community gathered at the cemetery with their picks and shovels. My dad said two men worked in the grave at a time. When they were tired, two more took their place. It was hard soil with lots of rocks and it is very difficult to dig a grave by hand.

My dad said when they were digging a grave if they found a part of a casket, maybe a handle which would not decay very fast, they would stop digging, refill the grave, and pick a new spot and start over. The reason they did this is because the cemetery is very old with some graves dating back to the early 1800s and many had no markers.

Neighbors also helped each other. If there was sickness or the birth of a child the neighbors came to help out. If the husband became ill and was not able to harvest his crop, the neighbors harvested it for him. It was the same with a death in the family. Nearly everyone in the community was no better off than we were and some had less. Everyone knew they might be the one who needed help the next time.

One summer I remember my brother-in-law was bit by a water moccasin when he was walking through one of his fields. He managed to get back to his house but he almost lost his leg and could not do any of his work to harvest his crop. The friends and relatives in the community pitched in and harvested it for him.

KENNETH GUY LACY

My parents were good to help neighbors in sickness and death. They would sit up with those at night. My mother helped in the preparation for burials. If the deceased was a female, Mom would go to the family's house to help.

If someone had not already closed the person's eyes, she would close them and lay a coin on each eye to keep them closed. The size of the coin used would depend on the size of the person's eyes. A grown person would probably need a silver dollar, whereas a child might only need a quarter. She then would bathe the body, put on the burial dress, the choice of the deceased or that of some family member, and help "lay the body out."

If the coffin was there, she would line it with some soft fabric before putting the body inside. Coffins were usually hand-made by a local caprenter. If the carpenter had not finished the coffin, someone would help Mom lay the body on boards until the coffin arrived. Sometimes if there were small children, she would bring them to our home to help out the bereaved family. I remember that after one funeral, the entire family—parents, two boys, and five girls—all came to our house and spent the night. We must have had pallets all over the place.

EVELYN LANGLEY

We went to church in the wagon. Of a night, Mama would put quilts down in the wagon. When we'd come home, we'd lay down and go to sleep. But it was three miles to the church and it took us a good bit to come in from home. We had all-day singings the fourth Sunday at New Hope Baptist Church in May every year. Everybody took dinner and got out on the ground, spread out our table cloths, and put our dinner on 'em. We passed out plates—we didn't have paper plates back then. The women had to take plates. They'd pack their dinner in tubs or boxes. Every woman took a plate, spoons, and forks for every mouth she had in her family.

DULAS L. MASSEY

Everyone was poor and we all helped each other. No house doors were locked. Church was an important part of our lives. We went to Sunday School and church every Sunday. In those days, we had just one church—a Methodist church. Later on we built a Baptist church. Now I think they have two of them but there aren't any people down there anymore. I still have my old family Bible right here by me on the table. We learned about Christ and everything, and knew the difference between hell and heaven. The church was the main thing in the community. Once a year, we had Decoration Day (of cemetery) and dinner on the ground.

EVELYN M. COONFIELD METCALF

Good neighbors. If there was an illness in a family, one of the older children walked or rode horseback to tell us of their need. We had no telephone and few cars, so everything was done by word of mouth. The ladies prepared food and the men did the chores, if needed.

If there was a death in the family, we would take food, serve the food, clean up, and help wherever needed. We didn't go to the funeral home like we do today. The body was bathed, clothed, and placed in a spare room or bedroom. I remember my mother and grandmother and other ladies sat around the clock, bathing the parts of the body that would be seen with rubbing alcohol to preserve the body through the funeral. Coffins were hand built, graves were dug by hand, usually within one and one-half days after the death occurred.

EVA SMOKE WELLS

We walked about two miles to church. We left home, walked down the hill to get on the highway that led to the church. When the revival came along, daddy always went with us. He would take a bunch of splinters and hide them in a certain place. [The splinters, Mrs. Wells explained, "were taken from pine stumps, which, after turning to pitch after the tree is cut, burn nicely."]

And when we come out of church it was dark. You didn't have flashlights back in those days. They were too expensive. So he would get to where he left that pile of splinters. He'd light 'em and he had a torch so we could halfway stay on the trail coming back to where we got to the road. Daddy made sure we went to church, which was the Magnet Cove Missionary Baptist Church. I've been a member there since 1938. We didn't have any close neighbors, so it was really a joy to go to church and see all the people who came.

ARWILDA WHITESIDE

We prayed together every day, and all the neighbors helped each other. Everybody in the neighborhood was in the same condition. My daddy had an old grocery store building. He would always buy extra groceries—extra peas, extra pumpkins, other things in that store, and when things got rough, people would come and say, "Mr. Paul, can we have such-and-such stuff?" and my daddy would say, "Yeah!" My husband's mother could sew—she could take the sacks we used for picking cotton when they got old. She could make the boys their suits and you wouldn't believe [how good they were]. So far as our faith was concerned, we just kept it. You can't explain it.

CHARLES WHITFIELD

Yeah, we went some to the Bismarck Baptist Church up on the highway. Church was a comfort to us, helped us get through some hard times. We had to walk practically every time they had church. Before we moved here, we was living up north on another forty and we would walk to church from there. When we went, we had to walk, you know. The preacher would come very seldom. They was paid hardly anything. Most of them would come by riding a horse and or coming in a buggy.

Back then, seems like people took Christianity and the church a lot more serious than they do today. They'd have revival meetings, and Lord 'a mercy, there would be no tellin' how many people would come. You'd have a house full. I remember a preacher named Chastain coming up many years ago from Texarkana. I'll bet he baptized twenty-five or thirty. Now you don't never see nobody being baptized. *Any advice?* Well, one thing, never be a crook. Pay your debts, if you can. If you can't, don't deny them.

LUCILLE RIDER WILSON

I can't remember a time in my life when I didn't pray. I depended on God for everything. I talked about everything with God. He was my confidant. My parents were not religious. I went to church (walked) but my parents did not. Between laying by and harvesting the crops, there were times where we would go to summer revivals that were going on in surrounding communities by different church denominations. Church was about the only place to go. I rode the school bus twelve miles to school. So we could walk two and a half miles to church. Sometimes people would be driving and would pick us up.

Once a year the church did the all-day-preaching and dinner on the ground. In the spring, every cemetery would have kind of a "Decoration Day." They were set up on the first, second, third, and fourth Sundays in May. It was like a reunion with a potluck on improvised tables and lots of food choices. We usually went to all the Decoration Days, wearing new dresses and summer shoes.

It wasn't a clean-up but more like a social event, a potluck dinner. People from away would come back at that point in time. The church would put potluck dishes on the improvised tables that would be set up. It was an all-day affair. We would have services in the morning, then we would eat the lunch and have some more services in the afternoon. Even though the reunions were in the cemeteries, I don't recall any of the ceremonies being around the tombstones. It was a big event.

Church was our social life. That's where my friends were. We went to church on Sunday morning and Sunday night if we didn't have to work the next day. If we had to work in the fields the next day, Daddy didn't want us to go to church because he wanted us to get our rest. It was a very small Baptist church, with thirty people max, and a lay preacher. Just Sunday School and church. Seventy-plus years later, I captured these memories in the following poem:

After The Big Meeting

On the way home
Old Ruth, Betty and Black Beauty walked sluggishly pulling
the farm wagon down State Road 12
Mama and Daddy sit on the spring seat chatting

about the revival sermon . . . and the neighbors.
I catch a word now and then
as I lie in back on a tattered quilt,
gazing at the stars and moon in the sky
contemplating the mysteries of God's hidden universe.
For just a moment He touched me,
And I felt the awe of being His.

Lou Wilson
July 2000 • Fairfield Bay

FLOYE WINGFIELD

My folks were quiet about their faith. They thought you lived instead of told it. I can't begin to tell nobody exactly how I feel about my faith and what my Father in heaven has done for me. We went to church every Sunday. We would have people who would come to the dinner on the ground, maybe walk ten miles to get there. But they always came with deep pockets in their coats. They filled them up with food before they left because they didn't have any food to eat at home. And it was worth walking the ten miles for.

What kind of preaching did you hear back in those days? Well, it was a little more fire and brimstone. I think preachers had a little more "my way or the highway" attitude. "If you don't believe as I do," they would say, "then you're going to go to hell anyway, so no need to come up here and bother me! I can't do much for you." With that attitude, the preacher couldn't do anything for me. I didn't have to listen to what he said.

JACK WOOD

We went to church every Sunday at Mount Bethel, out in the country. After church we would all have "Dinner on the Ground." All the ladies would bring the food and each family would put white sheets on the ground to spread the food on. We'd go up one side and down the other and that was one good meal we got a week. We had a lot of fun, you know? The only drawback was getting in our wagons and going the two and a half miles back home. We had no cars to travel in. The church parking lot had only wagons and mules and a few horses. I never did know who cleaned up the lot the next day.

[Author's note: From interview conducted by Maggie King, Jack Wood's granddaughter.]

Chapter 24

The Most Serious Problems That Were Faced

BESSIE BLACKNALL

Lack of money was our most serious problem. We couldn't get no meal or stuff. We had a gristmill if you had corn. We ate a lot of corn bread as I was growing up. I remember one time we didn't have nothing to make corn bread. Sometimes, we would put the corn bread in water, put it on the stove, put salt and pepper in it, and stir it up to make it thick. We called it "gruel."

FRED NORMAN BLANKENSHIP

The most serious things we faced were drought, insects, and boll worms. Drought probably was the worst thing because there was one year in the early 1930s when we didn't pick a bit of cotton that year. That year, the worms got it. We raised just about everything we would eat. So we had plenty to eat all the time and had as good a clothes as other people did have. The next year, we poisoned the boll weevils. We put some chemicals into two flour sacks, walked between the rows, and dusted the cotton.

Lack of money was not a problem because we didn't have anything to spend it for. If we had a picnic, daddy would usually give me fifty cents. I'd get lemonade for a nickel, a box of Crackerjacks (with a gift in it!) for a nickel. At the end of the picnic, if there was any lemonade left, we could have all we could drink.

LETA TISDALE CURRY

Lack of money. Daddy lost our farm in the mid-1930s because he could not pay the taxes. Of course, my parents were very upset losing their home but were fortunate to have a daughter who owned a small farm for us to live on.

WILLIAM EDWARD DELAMAR

Well, things was pretty tough, pretty tough. I can remember when we just barely had food. Of course, many people—black and white—had the same problem. We had to learn to help one another. If it hadn't been for people helping one another, I don't know how we would have made it. There would be times when I would have and you wouldn't have. Times that you would have and I didn't have. That's the way we did it. When I was young, I couldn't see much difference between anyone.

S. LORETTA ECHOLS

I think the most serious problem was the fact that there were not jobs available that would pay you enough. My dad always had some sort of work, but he didn't always get paid in cash for it. You might get paid with some goods that the family needed. I do know that early on he went to work in some of the WPA [projects]. I also remember an uncle that was just a teenager and he went to the CCC camps and went off and [the reason] that was just a big deal to us is that he was experiencing FDR's New Deal plan. I know for sure the thirties, even 1938–1939 were such hard times that it was often difficult for families to keep their optimism. We were asking, "Is it always gonna be this way?"

GENEVA KING EMERSON

Lack of money. Not having an income affected everything in life. Nobody went to the doctor unless there was sure to be lifelong consequences or death was thought to be imminent. When someone sent for the doctor, who had to come horseback, usually it wasn't known how they would ever pay the bill. Doctors waited for pay. To the patients' embarrassment as well as the doctor's trial of patience, it might take years before the full bill was paid, maybe in farm produce or animals.

Doctors had hard times like everyone else, and usually had a lot of compassion, giving out advice to help families treat their sick themselves. Babies were born at home with "Granny Women" usually volunteering their time. Their pay was often having the baby named for them. Perhaps they received some produce or specially baked foods or cooked meals—the very best the new mother could prepare when she was "back on her feet."

I remember one of my sisters was born with an umbilical hernia, somewhat worse than what was expected when this occurred. In fact, she was quite ill otherwise. All the extended family members were worried and there was a very sad feeling which I sensed. Maybe I was going to lose my little sister. Being about four years old, I don't know that what the doctor decided was wrong, but I remember he left instructions for making a lead weight for the hernia and some terrible-smelling medicine which had to be purchased and administered for a long time. I don't know how my parents ever paid the bills for all that. Probably, Dad had to sell one of our few farm animals.

ZELMA LOUISE HAMILTON

Lack of money. At one time, Daddy had been quite wealthy but because he was too generous, he gave away so many groceries—big sacks of flour, sugar, meal, and fifty-pound cans of lard or shortening, whatever they could eat—without being paid back. He lost almost all of his money but never complained. The blacks chopped briars on his place to pay him back, but they couldn't pay him back in money because they had no way to make it. Daddy never regretted doing what he did. However, he was just happy he could help.

Bank foreclosures. A friend "tipped" Daddy off two days before the actual closing. He took his money out, which was quite a "nest egg!" Where he put it, we, the children, never did know. Many people went around murmuring, "Ten-cent cotton and forty-cent meat, how in the world can a poor man eat?" These people lost all their savings when the bank closed.

Sickness. This was a serious problem because of transportation—just a wagon to get you to the doctor. My half-brother had appendicitis. The doctor thought it was double pneumonia until his appendix ruptured. He died a few days before he was to get married. My half-sister died of congestive heart failure. I also had two half-brothers die—one died as an infant, the other was about four years old. No way to get them to a hospital. I had a terrible case of chicken pox. My mother told me that she couldn't touch me anywhere for so many sores. She said that she held me on a pillow.

MARY FRANCES LOVELL IZARD

The most serious problem for us was a lack of cash money. My dad told me years later that everything was cheap, but there just wasn't any money and no jobs where you could earn some. Lots of people moved away hoping to find a job. Before the Depression, the hills in our community were filled with small farms. A few years into the Depression the hills were filled with abandoned home sites.

One uncle made bootleg whiskey. One of my dad's favorite stories was that my uncle had a still made of solid copper. In fact, when the revenuers found it they didn't chop it up but carried it back to town with them.

Losing the crop was a major disaster because it meant no income for the year. When this happened my dad tried to find outside work. He usually worked cutting logs with a crosscut saw for about a dollar a day. One year he learned that the Cache River bottoms near Bald Knob were being cleared. One neighbor owned a Model-A. They packed all the men they could in the car, split the cost of gas, and headed there to make some money. There were large lumber camps that hired the men. The men were provided with tents to sleep in, and there was a large "cook tent" where a cook prepared their meals and in which they ate. The lumber companies were cutting bolts, a short size used to make barrel staves and the men were paid by the piece. It was not a lot of money but it was a job when jobs were few. The men stayed there most of the winter.

Everyone in the family made money any way they could. There was no market for eggs, butter, or produce because everyone else lived like we did. After our cotton was laid by, my dad and the older girls would chop cotton for other farmers. My sister said they worked from dawn to dusk for one dollar and twenty-five cents a day. The men also hunted opossum and raccoon. My dad kept special boards on which he stretched their hides and after they cured, the hides should be sold for about fifty cents a piece. Everything was used, nothing was wasted.

WILLIS MAGBY

When I talk to people who have not gone through something like the Depression, I don't expect them to believe it. And you know what I would say to those who expect a second "Great Depression"? It would be like a Sunday School picnic compared to what we have now. It was rough. We went from here to Texas in a Model-T Ford in 1933 without hardly any money. There were nine of us— Mom and Dad and seven of us kids. I was the oldest at thirteen. It was the last Model-T that was made in 1927, so it was only six years old.

Those cars wore out right quick. My dad had a brother down in south Texas. Cotton was booming around here and down there they was pickin', you know. So we sold everything we had and went to south Texas. We was farming and had a crop but we sold it to a neighbor. It was a rough time because we didn't have hardly any money at all. So when we stopped along the way at night, we just slept on the ground. One night it was raining and we didn't have money to spare to go to a tourist court, which would have cost a dollar and fifty cents, probably. So we slept on the ground.

Another time during the trip, we were just about to run out of gas. It only cost seventeen cents a gallon, but we had also just about run out of money— I think Dad had thirty-five or forty cents. Dad had a nice Smith & Wesson .38 pistol. He said, "We'll go as far as we can." When we stopped at Wharton down in south Texas, he pulled into a station and said, "We'll go as far as we can." A tank of gas would get us on to my uncle's house in Taft, Texas. When he asked to pawn that gun, the station owner asked him what he wanted on it. Dad said, "A quart of oil and a tank of gas." It was a dollar forty-five. I still remember exactly. That's what it would take to get us from Wharton on to Taft, which was another hundred miles. And in a Model-T, that was like eternity.

"Is there anything else you want?" the owner asked. "No," Dad answered. The owner wanted the gun. It was beautiful. I'd give a hundred for it now. Dad asked the owner for his name and address so he could pay him back. The station manager said okay. We went to picking cotton as soon as we got to Taft and Dad paid him back. When the owner received the money from Dad, he mailed the pistol back to us.

ARWILDA WHITESIDE and CLEOVIS WHITESIDE

Lack of money. ARWILDA: They had stores in Clarendon—dry-goods clothing stores, grocery stores. Back in those days they would allow you fifteen dollars a month. You could go to the store and get you fifteen dollars' worth of groceries. Or go to the other store and get your clothes. Then when crop time came, that's how you paid them back.

Lack of medical care. Not like it is now. People didn't get sick all the time. If they did, we lived in a community where you prayed, used herbs. If you got too sick, you died! [laughing] But that's the truth!

CLEOVIS: That's right! If the doctor didn't get to you on time, you just died. We didn't have any hospital to run to no how. It wasn't like it is now. Every time you stump your toe you go to a doctor. But back in those days, people got together and prayed.

CHARLES WHITFIELD

Mama and Daddy didn't know what was coming next. But we always had something to eat. I remember one time when I had a pair of little gray mules I farmed with and I had a wagon. I had to get them mules some feed before I could get a crop laid by. I was buying most of my groceries in Marcus, which was about five miles away. So I went up there to get me a sack or two of corn.

They couldn't sell me anything on credit anyway. He told me, "Charlie, I ain't got a sack of corn in the house." He said, "Lee Dickson's got some across the street. Go on over there and get you a sack." I said, "I ain't got any money to pay it." And he said, "He'll let you have it." So I went across the street and asked him if he would sell me some corn on credit, and he said, "No." Well, I come on back and he said, "Did you get some corn?" and I said, "No, he wouldn't let me have it." He cussed right big—I won't tell you what he said. He jerked out his cashier drawer and let me have the money I needed. So I had the corn to feed my mules after workin' them.

Chapter 25

Worst Memories

KATHRYN BAJOREK
We had three rooms, no running water, no toilet. We had to hide in the field to use the bathroom. We would be so cold in the winter. At night Mother put a hot iron at our feet to help us go to sleep. She also heated water and would put it into gallon jugs. But there were just so many irons.

LOUISE J. BRANT
The flood of 1927 reached our home but not our farmland. Before the water got up to the house, we could stand on our porch and see uprooted trees and brush going down the river. So we had to move over to a neighbor's house and stayed over there two or three days at least. They were good people, but they had a large family, too, but they were willing to share with us so we just managed the best way we could until we could get back home.

When we came back home, the water had gotten up to the mattresses. Mother said there was even some ice left on the matresses. Neighbors came in and helped my mother gather up laundry. I remember two yards of solid red fabric that my favorite uncle had given me to make a red dress because I was a red head and my mother didn't want me to wear red. But Uncle Otis said that if a kid wanted to wear red, she ought to be able to wear it. I remember when they hung out the laundry that my two yards of red fabric were hanging on the clothesline. [chuckling]

R. L. "BILL" CARTER
My mother lost some money around 1929 at a Brinkley bank and never did get it back. Daddy didn't have no money. He used to borrow a hundred dollars to make a crop on. Before he started borrowing a hundred dollars, he traded at the Vernon Williams store in Monroe. Vernon didn't have no education, but he was a smart old man. He knew what everybody got and he just let 'em have it.

They used the honor system and they would pay him back. Finally, Dad decided he would just borrow money from somebody, and he'd borrow one hundred dollars to make a crop on. That had to last him from the first of March until we got the first bale of cotton out in the middle of September. He would go into town, buy a two-hundred-pound wooden barrel of flour, a hundred pounds of brown beans and a hundred pounds of white beans, and a hundred pounds of sugar and a stand of lard. And that had to last until we got our first bale of cotton. And you didn't throw that flour away just because weevils had gotten into it. You'd take your sifter and sift them out and eat it anyway.

MARY LEONA CARTILLAR

Nothing to eat! I mean, when I say nothing, I mean nothing! One year we raised potatoes. And, we had a lot of potatoes that we boiled, and all we had to put on them was salt. We didn't have no other seasoning whatsoever—no bread, either. I well remember all of that. We didn't have anything to eat. There was no money. There was no way to make money. I well remember that there was no food and no way to get any. So all you had was what you raised.

S. LORETTA ECHOLS

Kerosene mistaken for water leads to death. James Sidney, my little brother, died in an accident at home when he was only eighteen months old. We didn't have a lot of things that are safer today than they were then. We had kerosene stoves, woodstoves, and all that, but for our cookstove, we had to use kerosene in it. And my mom and dad had a system that failed. They had a jar of kerosene sitting up on a shelf in the kitchen to use to light a fire, but we used fruit jars to put the kerosene in. Well, we didn't have separate glasses. We drank out of fruit jars, you know. And so we had the well water and all of that.

But anyway, one day my sister thought there was water in that fruit jar and gave it to my brother to drink. Now if we had lived in the city or if times had been different, if there had been, you know, more money, we may have been able to take him to a hospital and he may have lived, you know, but there was nothing we knew to do. Word got out and I do know now that the doctor came out but my brother did not survive.

JEAN EDWARDS

Although we had food, we had friends seventeen miles away who did not have any food. So we walked to their home and carried corn for them to grind and make meal for corn bread. They were good friends. The way we became acquainted with them is that my father had rented to them a part of a duplex.

They lived on one side, we lived on the other. After we returned home, we still kept them in our memory because their father lost his job. They were suffering. We had corn because we had mules and hogs that we had to feed. So we carried the corn to them and they used it to make hominy and cornmeal out of it.

GENEVA KING EMERSON

The hell of summer heat, humidity, and field dirt with no relief until we could finish the day's work and get to the spring branch or a pan of water. Our country school dismissed early in the year so the children could help with clearing the stubborn, re-sprouting brush from the fields, planting and working crops, picking strawberries, making the garden, etc. I found it rather enjoyable until about the end of May when the summer heat began to pick up. Soon it began to feel nearly unbearable.

We did chores at daylight—milking, feeding chickens and penned animals, carrying water, getting in wood for the cookstove, washing the breakfast dishes and pans—and we made sure we ate a big breakfast, because you did not want to get caught in a field a long walk from home and just have to keep working when we thought we'd pass out from hunger.

When dinner time came ("lunch" today), the effects of the heat and being so tired from work often made it difficult to eat a big meal even when we were hungry, but eat we did and lots of it because it would be a long time until supper. There was no electricity for fans to cool us. Just dipping our hands into the washbasin and splashing our heat-reddened faces and arms was such a luxury! I'd hold first one foot and leg, then the other, out over a flowerbed by the porch and pour my used water from the wash pan over it while I thought of the "Rich Man in Hell" Bible story. I tell you: when our pastor preached hell fire, we didn't have any trouble understanding the concept!

After dinner, we carried more water and did any other midday chore that couldn't wait. Then we took a jug or bucket of fresh drinking water to the field, maybe tucked a leftover piece of bread or 'tater, if there was any, in our pocket and went back to work. Mom packed a box or old quilt along to keep the baby out of the dirt and in the shade of a tree. We helped her keep an eye on the little one while she also labored in the heat and humidity.

Those hoped-for harvests were our living and we all had to help make it. While chopping cotton, we'd use our hoe to pull the heated topsoil from a small place in the middle of the cotton rows so we could have a cooler place to place our bare feet. When we'd hoed as far along the row as possible from that spot, we dug another one and moved to it. Sometimes I could find a white rock to stand on; maybe move it along ahead of me. It seemed cooler than the ground.

What a relief to see the sun getting low. Some days, a little gentle breeze would find us, and we knew God loved us. But there might be many days in a row—even weeks—without a breath of air, and we and the crops burning up, then we wondered. I guess our faith got a little weak, at times. No matter what came, we knew we must persevere; there was no acceptable alternative.

At day's end, how wonderful it would have been to climb into a tub of cool water or go to the spring branch and lie down in it. We sometimes managed the latter for a few minutes if we could get through the chores in time. I treasured any little stream I could cross as I was bringing the cows home for milking. I stole just a extra minute or two to wade in it a ways and splash the water on as much of me as I could. By midsummer most streams were dry. We managed to quickly dip the poor baby and little children in a pan of water or rag-wash them. We joked about the rings of dirt in the folds of the little necks: "Here," we would say, "let me wash off your beads!"

Finally, before we dropped onto a shuck or straw bed, or to a pallet on the floor, we applied the washrag and got on our clean sleeping garment (I knew neighbor kids without that luxury). It wasn't unusual for the nights to be so hot, we couldn't sleep except from pure exhaustion. I found one comforting thought: The watermelons would be so very sweet. Hot nights were good for them.

By daylight, it all began again. We were glad for the beginning of the late-summer-school term where we could be in the shade and study. Our sweaty arms sticking to the paper and pulling it up off our desks hardly bothered us. With the crops laid by, it was time for revival meetings, too, and we were happy to gather with our neighbors to sing and listen to sermons in hot buildings, much preferred to working the fields under the boiling sun. Then came the end of the summer school term and the beginning of fall harvests, which were started in the heat and often finished in freezing cold before the winter term of school could be started.

ROBERT EUBANKS

Probably the uncertainty of the future was our biggest concern. You'd listen to the adults as they talked politics. Back in those days, the Democratic Party was the one almost everybody gave allegiance to. They stood for the same principles of today's Republican Party. The Democratic line has gone way liberal to what it once was.

LUCILLE KING HAVNER FINLEY

I was in my twenties, and married, when the Depression came. I remember well what a hard time we had. We were hungry, and we didn't have nothing hardly to eat, and couldn't get it. My husband was disabled and couldn't work. Me and the kids just had to manage the best way we could. One day I set out on foot to Calico Rock to try to get some flour. My first stop, after walking five miles, was one of the main grocery stores in Calico Rock. I asked if I could buy 25 lbs. of flour on credit, and the man said, "No." I was very disheartened, but still very proud, and knowing that I had to find someone who would let me buy 25 lbs. of flour on the credit so my family could eat.

Weary and very heavy-hearted, I started on my way home. Not one to give up easily in time of despair, I decided to stop at the store of William and Cecil Jennings. I went in and asked if I could buy 25 lbs. of flour on the credit and they said, "By ★★★, sure you can." They treated me with utmost respect and I carried the 25 lb. sack of flour the five miles back home to my family. Now we would have bread to eat.

Adolph sold his coon and possum hides that fall, and the first thing we did was pay the store for the flour. Neighbors were good to each other, but all of 'em was in the same boat. Didn't none of 'em have anything hardly; they's all in the same shape. I wouldn't ever like to see a time like that again.

[Author's note: Excerpt from "Down Memory Lane." Used with permission.]

MARY EATHEL FREEMAN

Not having enough water on the land Dad owned, seeing him worried about getting debts paid, sicknesses. The year I started the ninth grade, my dad was hurt in a logging accient. A broken branch came back and hit his collarbone and cracked two ribs. There was no hospital insurance in the thirties. We went to the doctor, and was sent home.

Frank, my younger brother, dropped out of school and took my dad's place in the logwoods. I dropped out and stayed at home to care for him. It was a hard time financially. We wore old clothes and ate food produced at home. I had missed six weeks of school before my dad was able to care for himself.

During the early thirties, Mama had a nervous breakdown. She had lived a life of poverty in her early childhood. Her mother had died when she was six years old, and her stepmother was crippled. She had bore four children and nursed us through sickness. She had seen my dad through typhoid and malaria fever. Combined with the economy, things must have seemed hopeless. But

Mama was a devout Christian, and that may have helped her through the terrible times.

[Author's note: Excerpt from Eathel Freeman, *Journey of a Depression Kid,* 3rd printing (self-published, 2005), 17, 18. Used with permission.]

VERNON L. GLENN
We were just barely getting by and we didn't have any money. When we would go to school we didn't have any money, like the kids now think you have to have money. I remember, for example, a big thing at Lynn—we lived about two miles from there—there was a Fourth of July picnic. About forty or fifty people would come in and you'd get to see all the kids, but I remember my birthday was June 29—and I will always remember this—my dad gave me a quarter. He said, "I'm gonna give you two dimes and a nickel and you need to hold on to this for the Fourth of July picnic." My mother wrapped it up in a handkerchief and we put it away.

This was June and on the fourth of July my dad comes back around and says, "I really hate this, but I need to get fifteen cents that we gave you." Austin, my older brother, was about sixteen then and he was kinda looking at the girls, so my dad didn't have any kind of money except the money they gave me, so he took fifteen cents and gave to my brother and I still had a dime. But I was still happy with that 'cause I can tell you what I spent it for. I spent it for an ice cream cone—you could get it for a nickel, and one of those little boxes of popcorn, Cracker Jacks. That was a great day for me.

VIOLET HENSLEY
We wanted to go to a carnival. Another bad memory is the time when there was carnival in our community. My two sisters and I wanted to go see it but my dad said, "No." He'd seen them and "there was nothing to it," he said. But since we had never seen one, we couldn't know that. That made us real sad.

DON ORVILLE KNOLL
Daddy didn't believe in women being able to write a check on her husband. He gave Mother five dollars per week for groceries. She would try to save a little of it for us to go to the movie on. Saturday night was a movie night.

WILLIS MAGBY
The winter of 1935–36. The worst came first. In late January or early

February 1936 it came a snow where we were farming in the Red River bottoms in Miller County. We couldn't find a day's work because there wasn't much going on. We had very little to eat for several weeks. A man who owned a store gave us a little bit of stuff during the wintertime. My brother-in-law and me went out in the fields, tracked up, and killed eight rabbits. It was the first meat we had in several weeks. We cooked and ate all of them before we stopped, the ten of us.

We borrowed money from the government to make a crop on and bought a team on the twenty-eighth of February, I believe it was. And as part of the loan, we could start buying some groceries. We had a few chickens that we sold and bought something to eat from that. It was bad! I can't even tell you how bad it was. Nobody would give you fifty cents a day to work because in February and March, there was nothin' to do! And even if the farmers needed hired hands, they got 'em already.

DULAS L. MASSEY

When my brother died. He broke his arm in school and it was always crooked. When I come out of the CCC camp he was about thirteen years old and wanted to do something about his arm because it was always hurting him. We took him down to Morrilton and a doctor was going to reset his arm. The doctor there in Scotland didn't know anything about that. They operated on his arm, but he got a blood clot that went to his brain and that killed him. He was the only one of us that had any sense. He just sat around and read all the time. He would read every book he could get. He was only thirteen years old when he died.

MAMIE H. MAYS

The death of my mother. I knew she was sick but I didn't realize she was that sick. It was awful! My mama had a stroke when she was at the wash pot punching clothes. My little brother, who was eight, had stayed at home with her and the baby. My older brother was an overseer in the county. Me and my brother was at school, so they come and got us. When she had the stroke, she couldn't talk but she made my little brother know he needed to go get my neighbor, who was my daddy's niece. Mama lived five days after that stroke. It was in February when it was so cold!

She had always told me that when she died, I could have her gold wedding band. When I would come to the side of her bed, she would make motions to ask me to rub the ring. I thought she was paralyzed. It didn't come to me until after she died that she wanted me to have that golden ring off of her finger and give it to me.

Did you ever get the ring? No, they left it on her when she was buried.

VIOLET PIERCY

California memory. This was the only time Bill ever saw his father cry as he watched his family leaving for California in 1934 where they remained. Three became electricians, one also a plumber. We were the typical *Grapes of Wrath* family in California. We arrived in an old cut-off '34 Chevrolet with a wooden flatbed on which my father put benches across both sides and the front, and a rumble seat, which they covered with an awning. The two boys rode the rumble seat, and the others rode in front.

There were nine of us who went to California in that vehicle. They called us Arkies, Okies, and hillbillies. My mother made every stitch I wore. I had to wear the same dress to school all week long. The California families had more money than we did. I envied girls who had wool skirts and sweaters. My dresses were made out of printed flour and feed sacks. The fifty-pound sacks of flour were made out of pretty print. Mother would buy enough sacks with the same print to make dresses for all four sisters.

ALMA POUNDS

My father and grandfather lost all of their land when their bank failed. What I remember was the talk I heard about it. When Daddy had a lot of cotton, he was waiting to sell it until the price got higher. The banks must have closed before he sold it and we lost it all. The government came in and sold the farm to somebody in Jonesboro.

The day we moved ended up being the awfulest day I have every lived. It was the day we swapped houses because we had just lost the land. We moved to a house that was just north of where we lived. It was the first time I had ever moved in my life. I begged and begged to stay at home from school that day, so Mother and Daddy decided to let me stay. The people [who got our house and land from the government] were going to move into our house and we were moving into theirs. At first, I thought that was such an adventure just to live in a new house. But it was raining and the furniture was all over the place and I was so heart sick. I was so tired I wish we had never left.

I guess that's when it really hit me that we were in the Depression. I was fourteen so it had to be around 1928, the beginning of the Depression. I can't even think about what my mother must have been thinking. We were swapping houses! The house we had been living in was tight and warm. The one we moved into had cracks in the walls. It was just like a barn. In the wintertime, it was hard to keep warm. We had a coal stove. Daddy would get up and build a fire in the morning.

The ones who were big enough would go out and milk the cows and then come back in for breakfast and get ready for school. It was cold in the house all the time. I can remember cracking the ice on the water bucket we brought in the night before.

I lost a little brother in about 1935 during the Depression. We think it was polio. It was at the beginning of the polio scare. We took him to Jonesboro to the doctor. They put him in the hospital and he died later that evening. That was a bad memory. How did we survive? We just didn't think about it. It was just something we had to do.

UNION H. STOUDAMIRE

My worst memory was when the tornado came and killed my great-great-grandmother. It was in the thirties. I was just a girl when that tornado come on through there and broke my grandmother's arm. There wasn't nobody at home except me and my two cousins. We could see that tornado going along the other side of our field up the Arkansas River. Trees and clothes and things were up in the air. Somebody said someone's player piano had blowed out and it was going through the air turning over and over and playing, "Nearer My God to Thee." I don't know if that's true but everybody said it was because that wind got in it and made it play that music. But I was scared that evening, I'll tell you. I learned to pray right then and I've been knowing how to pray ever since. [laughing as she remembers]

EVA SMOKE WELLS

The only thing that ever bothered me was going to school. My mother did not sew. She made quilts but she could not make me a dress. My grandmother wasn't with us all the time. She could have made a dress, but she was a practical nurse and had to take care of other people. Before school started, one time I went with my brother over to the Meadowview community—it's no longer there. This friend had a daughter who was in the same grade with me in school. She couldn't wait to get us in her room so she could show us her dresses and all the materials that her mother used to made her dresses. That was one time I was envious, because I knew that Mother couldn't sew and she wasn't going to make me any dresses.

LUCILLE RIDER WILSON
Somewhere down the line, Daddy had borrowed some money from somebody named Johnny Baker. I look back on my childhood and it seems like every year Daddy was paying Johnny Baker. When the bank at Ratcliff went broke, Daddy had borrowed money from the bank. I think they transferred his loan over to John Baker. That's what I think happened. A long time later, I asked Daddy how much he owed Jim Baker and he said, "Fifty dollars." But you know, fifty dollars was different then. Had Daddy been paying two dollars a year for years?

FLOYE WINGFIELD
We had a funeral, and this man Bud was so poor he couldn't even be a sharecropper. But that's the kind that marry and have children, and he did have a family. Bud died. The men in the community built a coffin out of lumber they happened to have. The women made a cloth lining and put it in the coffin. They were carrying him to the graveyard in a wagon and the wagon didn't have an end-gate and they had to go up a steep hill. His coffin rolled out and Bud rolled out, too.

I remember, oh, how I remember how horrible had it been some of my people. How did his family react to it? Bud's family didn't know the difference. I realized even then that I want some mourners to be at my funeral. If there's any money left, I want my family to buy me some mourners. My nephew volunteered to take care of that for me, and he will be punished if he doesn't do it. Just as sure as anything, God will punish him.

Chapter 26

Effects of the Agricultural Adjustment Act

R. L. "BILL" CARTER

I remember plowing up cotton that had already been laid by. The government would give us a little ol' check that didn't amount to much. If you had hogs, the government would tell you to kill one of them but not to eat any of the meat. Take it out in the wild and a government man would make sure you had killed the hog. We should have been able to have at least one good meal. The buzzards had a feast. I was only five years old at that time but thought how stupid can the government get?

DOYLE L. COLLINS

A sad time was also when farmers were required to sell cows to the government. Government representatives would take the cows to a central location and shoot them because there was too much milk in the country. This was the result of another government program. There was a surplus of milk in the whole country, and you couldn't sell it. So President Roosevelt established the Agricultural Adjustment Act in 1933 in which the government would buy some of those milk cows, take them to some remote place, and have them shot. They had a place that was about three quarters of a mile from our house where they arranged to take all those cows.

The farmers were allowed to select so many of their cows to sell to the government to decrease milk production. We selected our oldest cow, "Old Daisy," to be taken over there for disposal. I remember what a traumatic experience it was. It was a pretty day and I, as a little boy—I was about six or seven years old at that time—was out around the barn playing and I heard the shots. I remember thinking, "Was that the shot that killed Old Daisy?"

BEULAH LEE MCLEOD EVANS

Coming out of the Depression was most painful. The government forced us to kill our cattle and hogs down to a certain number. They checked on you to see that you did this. You couldn't eat the cattle and hogs you'd killed and you couldn't give them to the poor people that was starving to death. You had to kill them and bury them. Even now, I've never understood this, but some people said it was to bring the prices up. There's a place over here between Powhattan and the Three-Way Inn, out in that field they buried—oh, there's not any telling how many they killed. They brought bulldozers. They'd bring their cattle up, they'd dig big ditches and cover them up.

Do you know how many animals your daddy had to kill? Well, I don't remember how many, but I remember one of them was my cow that I had learned to milk. I started milking when I was nine years old. And I learned to milk "Old Nanny," and she was one of them. You could choose which ones, so they chose the ones that were the oldest and less valuable. We didn't have beef cattle; we raised dairy cattle and sold cream.

Did your daddy try to save some of the meat for canning? Oh no! He wouldn't have dared to do that! Daddy was a straight man. But I remember that the first time I saw my daddy cry was when he had to kill them cattle.

ROSCOE E. JEFFERSON

Government destroys cows. [We were ordered to take our cows] and sell 'em to the government. We didn't burn them. The government did. They had a feller who would knock the cows in the head. Then they'd pile 'em up, put gasoline on 'em, and set 'em afire. Why? Because we had so many cattle and they weren't bringing nothing. A cow, the best ones, would bring about twelve dollars.

[Author's note: For more on the Agricultural Adjustment Act, see page 22.]

GRACE FERGUSON JOHNSON

Another bad thing about the Depression, the government was making people get rid of part of their cattle, at least two head, and they had to kill 'em. They took 'em out on the other side of Mt. Pleasant on some of them hills back there. I guess I must have been married then, and I can remember my dad had a cow and calf, and he went along and helped drive 'em out. That was the silliest thing. Looks like they'd have made them kill 'em and can 'em. The government give them a little money for that, but it wasn't very much. The purpose was to get rid of a lot of cattle cause there wasn't feed for all of 'em.

[Author's note: Excerpt from "Down Memory Lane." Used with permission.]

LLOYD A. PERRY

When the New Deal came in, the government began to buy up livestock and slaughter them to raise the prices. The carcasses were dumped in pits and old mines, just wasted! We heard they were even dumped in the ocean along with tons of other food—nobody had any money to buy it, but it was against the law to sell the meat.

I saw a pretty milk cow and her fat calf bought up for fourteen dollars then shot, loaded on a truck, and hauled out to an old mine near Cave City where a lot of slaughtered animals were dumped. People were so needy. Although it was also against the law, there were desperate folks who went by night to the dumps and dressed out the best meat to feed their hungry families. By summertime, if there wasn't a drought, farmers had a little pasture for feed and some animals were in good shape. They just weren't worth anything. I tell you, it was enough to might near make a Republican outta me!

CHARLES WHITFIELD

Controls on crops. The government allowed us to plant just so much cotton. And the cotton got so cheap we couldn't hardly sell it. The government would allow us to plant just so many acres. They would cut you down so much percentage from what we had been a planting. They'd come and measure after you had planted it, and if you had too much, you had to plow it up. And they'd pay you a little bit but I don't remember how much. It wasn't very much.

Chapter 27

Racism during the Depression Era

BESSIE BLACKNALL

Did being white or black have any effect on how you were treated during the Great Depression? No, not really. It didn't make no difference if you were white or black. We all tried to help each other. If you had a ham bone and somebody needed it, you'd give it to them. We would chop cotton for 'em, whatever.

WILLIAM EDWARD DELAMAR

Were blacks and whites segregated in your school? Back in those segregated days, all the students were black in my school. We had to walk to school on the highway. When white children were riding school buses to their schools, we had to walk. I didn't know what a raincoat was. If it was raining, I just kept on walking. If the water was across the road, we would wade it. And if a bus come along, I had to step aside to keep the water from being splashed on me.

Were you treated well when you got to town on the wagon trips? Yes. I have to say we were. That's one thing I'm proud of. I've never been involved in trouble. I never have been treated too mean. I always knew my stall and that's the one I stayed in. I knew my place.

I try to be careful about using the words "black and white," because I learned there were some poor white folks, too. All those people back then were just alike. Now they are not just alike. A lot of us is mad if some white person goes out there and does something. Then you say, "Well, they are all just alike." I knew that was wrong. On my side, black people will go out there and do dirty things. I don't feel good about it, but don't say I am that way, see? "All of them black folks are the same." Excuse my expression, but that's a lie.

Did you experience what you considered to be injustice as you were growing up? Where I was raised at, black people were not allowed to vote. Like whenever I wanted to vote for a president or anyone else in government, no, no—we couldn't vote. What could we do? We couldn't do anything. Really, being a kid like I was, I didn't know too much about it. We just did what we were told to do. You was never allowed to do it no way, so you get used to that kind of stuff, but you're behind on all the things you need to know.

Looking back on those times, how does it make you feel? A lot of folks want to act up on that. I don't. Out of all the hard times, I consider myself today to being blessed. The times . . . the times have really changed since I've been here. But through all of these years, I never was bad about hating anybody. I have some words I use a lot of times, and I hope you understand: "If your dog bites, I'll stay out of your yard." So if I knew that this was the wrong place, I would stay out.

[Author's note: Asked to respond to Mr. Delamar's statement that blacks were not allowed to vote, Tom Dillard, head of Special Collections, University of Arkansas Libraries, said this: "Actually, there were a number of strategies employed to prevent blacks from voting. The poll tax weeded out huge numbers of poor whites and blacks. Also, blacks were not allowed to vote in the Democratic primary until after FDR died (that came during the McMath administration). Technically, blacks who had poll-tax receipts could vote in the general elections . . . but it was the Democratic primary that was the main election in Arkansas.

"I suspect the person you interviewed was simply *intimidated* into not voting. Many blacks (even those who had poll-tax receipts) never tried to vote, because they knew it would cause a ruckus and that would have potentially caused lots of problems for the people involved." Much of the pressure and intimidation was not readily apparent. For example, a black man or woman who attempted to vote might find themselves receiving lots of parking tickets, or maybe could not get welfare, or maybe a grandchild would be expelled from school. It was not a fair fight!]

LEONA STITH DUFFY

In "An American Family," Mrs. Duffy said she told her children that "racism was a reality in their lives. They went to black schools. They saw the differences between black and white living conditions." But she then made an observation that must have meant so much to them: Racism, she said, should not make them feel inferior. "It's what you are inside that counts. You don't have to reconcile yourself to the racial conditions here. If you don't like them, then leave when you have the opportunity and seek something better."

[Author's note: Excerpt from "An American Family" by Joseph P. Blank, *Reader's Digest,* July 1977, 109–10.]

[Author's note: The following interview was conducted in 2009 with Nancy Duffy Blount and Spencer Duffy, Mrs. Duffy's daughter and son. Mrs. Blount has served three terms in the Arkansas House of Representatives, and eight years as Lee County coroner, the first black elected to such a position in Arkansas. She and her husband own two funeral homes, one in Marianna and one in Forest City. Mr. Duffy retired in 1992 from the U.S. Department of Agriculture and the Environmental Protection Agency in Washington, D.C.]

DOWNS: What experiences do you recall regarding racism back in those days?

NANCY BLOUNT: A couple of things that I remember. In the cotton fields, there would always be the straw boss, the white males who were over the workers. While we chopped and picked cotton, the white boys and girls would be playing or be off to camp or riding around while we worked in the fields. And no matter how old you got, if you were black in the fields you were always called by your name—"boy" or "girl." You never became a man or a woman. So Dad was never called "Mr. Duffy." He was always "Duffy."

I remember one day a gentleman came to the house and I was outside hanging clothes on the line. He asked if "Duffy" was here and I asked him if he meant Mister Duffy? Then he warned me to make sure that I didn't make any trouble.

DOWNS: What did this motivate you to do?

NANCY BLOUNT: In those cotton fields and in that hot sun, I thought there has to be a better way to make a living than this. And I made a promise to the good Lord that if I got out of those cotton fields and had an opportunity to do better, I would dedicate my life to helping children. That's how I got here to Marianna teaching school—working in areas where children were growing up very much like me to show them that they didn't have to stay in the shape that they were in and that education is the key out of poverty and into a better life. And you can do anything you want to do regardless of the color of your skin or where you came from.

So we understood that there was definitely a difference in the way we were treated just because of the color of our skin and our not being able to go into those little fast-food places. We were served on the outside while the little white children could, of course, go inside and sit down and eat at the diners.

Another thing was that we always got books and typewriters and other materials after the white children had used them. Then they would send them over to our school. I didn't know what an electric typewriter was until I was grown up.

SPENCER DUFFY: I was always curious about learning and so I began to seriously start planning to find a way to make a better life for myself. So from then on I began to really focus on "What shall I do to escape this condition?" Education was the key. It was the one thing that was available to us and it was the thing that would allow us to gain the knowledge to deliver ourselves from the situation.

But racism had another effect on me—it was the fact that I realized that I could do a lot more than what I had been exposed to. After I found that science had so many opportunities, I just focused on it and began to channel my knowledge into an area where I wanted to work.

LaVERNE WILLIAMS FEASTER

Was there much in the way of racial discrimination? We didn't call it that. Whites operated in one way, blacks in another. On Saturdays, we used to go to Stevenson's Drug Store in Cotton Plant. White children could go in the front door, sit at the counter on stools, and eat their ice cream. We blacks had to go in the back door, get our ice cream, and go out on the street. Our parents would not let us walk on the street eating. It just was not good behavior. We waited until time to go home, got our ice cream, and ate it in the car as we drove home. I would tell Mama and Daddy that a white girl who lived near us was sitting in there eating her ice cream. "Why can't I?" I would ask. Mama would say, "What difference does it make? She is no better than you. That is just the way it is."

WILLIE MORRIS

When you were growing up, was being black a problem? No, no! It wasn't for us, 'cause you know what? The McCloys, who had only one child, who owned the plantation, built a shotgun schoolhouse on their own plantation and hired their own teacher, so that about five or six of us black boys could go to school with their white son instead of going to school off the plantation with other black kids in the rural community of Baxter, Arkansas. The commissary on the plantation had groceries and dry goods, things like that. They also had a gas pump. If you wanted a gallon of gas, you would pump it up to where the glass tank would say a gallon, then you would use the gas hose to put it in your gas tank.

ARWILDA WHITESIDE

Was being black in those days a problem? We weren't bothered too much about it because we stayed in our place, and weren't allowed to meddle, and we weren't allowed to dishonor folks. It was always, "Yes, ma'am" and "No, ma'am," and "Yes, sir" and "No sir" all the time. And even if they were younger than we were, we had to call them "Mister."

We could play with them and be friends with them—the white boys. When we were playing with them we could call them Jack or Robert or whatever their names were. But when they got older, you'd have to say Mr. and Mrs. to them. We were trained to honor them. We stayed in our place.

Chapter 28

Best Memories of the Depression

DOROTHY COX COSTON ASHLEY
The comfort and security we felt to be constantly surrounded by friends, neighbors, relatives, and classmates. I knew I could always count on them for anything, anytime, and anyplace. I didn't know anything about the outside world! This was my world. I've always enjoyed life. I learned early that if I was going to have any fun in life at all, I was going to have to enjoy what I was doing wherever I was. We had to pick our own fun on the farm. My parents were loving, hard working, and fun loving. They learned to enjoy wherever they were and whatever they were doing. It was a valuable lesson.

KATHRYN BAJOREK
When we got to go to sleep after a hard day's work. Also my daddy's faith, my mother's patterns. Both made things from nothing. My own job at four years old was to dig a hole lined with a toe sack. We put milk in the ground to keep it cool. Some people today say they are bored. They didn't live in our house! I chopped stovewood when I was five or six with an axe. Do you think my little grandkids who are twelve years old would do that? Cutting that stove wood and bringing it in was how we cooked all that sausage after we had killed all those hogs.

R. L. "BILL" CARTER
Our family. We worked hard, and after supper we'd go out on the old front porch and Daddy could play a guitar. He'd play the guitar and we'd all sing. And then we would sit there and listen to the whippoorwills holler. I got a letter somewhere that I had written to the editor telling about how it was back then. I had the verse of one of the songs we used to sing. I went on to tell them that today it was a far cry from what it used to be. I told 'em about the old front porch. Now all we got is a one-eyed monster—television—with Hollywood blurting out vulgar and profane language.

MARY LEONA CARTILLAR
I guess being with my friends Carrie, Verla, and Rennie. And when we let the

air out of the preacher's tires so he couldn't get to church the next morning. Me and Rennie crawled on our stomachs to get to the tires. We let the air out of all four tires. I think he was a Presbyterian.

DOYLE L. COLLINS

My first ice cream cone. I must have been six or seven years old because I remember it real well. The person who bought that first ice cream cone for me was Emory Langdon, who lived close to the CCC camp where my brother was stationed. A group of these CCC boys would hire him to bring them back sometimes on weekends. They would stay the weekend before going back to camp. Emory would either stay with us or spend time at our house. And he came to know us real well. This was about 1935 or '36.

One time Emory put a few of us into a pick-up truck that he brought his CCC boys back, and took us out to Ash Flat and bought each of us an ice cream cone. I was afraid to eat the cone, thought it was poison or something. I didn't want to take any chances, so finally I asked, "What do you do with that cone?" And my brother, who was two years older than I was—I don't know how he found out. Maybe he'd had one once before that I wasn't aware of or something. He said, "You eat it, stupid!" or something to that effect. I remember digging in on that cone and how it tasted—it sure did taste good!

Hymn hollerers. I remember fondly the way people sang a lot when they were working, mostly hymns. We had a fellow, Joe Short, half a mile up the road from us. When he was out at night, he would go up the road hollering out those hymns. I think they still do that in the Carolinas. Ol' Joe would be hollering out those hymns when I was in bed, such as "When the Role Is Called Up Yonder" and "The Land Where We'll Never Grow Old." They didn't pronounce the words at all. They just hollered. I don't know how to project a "hollerin'" sound but if you went outside and your mother was calling you in to dinner, she would cup her hands around her mouth and holler to let you know that lunch time or dinner was ready.

Sharing with my buddies. One of the upsides, believe it or not, if you had something, as simple as a ten- of eleven-year-old boy, any age level, really, all the way up to your parents' age in a sense. If I had a treat and I had a buddy, that buddy would get half of it. That was just the way we did things in the Ozarks. If I ever got the money to buy a candy bar, I'd break it in half and offer it to my buddy. And I offered him the longer half. I would take the shorter one. We were taught these things.

The leadership and courage of my parents. You can imagine how hard it was for Mom and Dad to pull our family up and go to California in order to try to provide a better life for us. Although they never talked to me about it, I think Mom and Dad had a sense of failure when they had to give up on that and come back to Arkansas. It did take leadership and courage to have the guts to go through a crop without the promise that you might make enough to buy your kids clothes for the school year.

My mother's steadfastness. My mother had a rough, rough life. Probably the roughest life of any member of our family, particularly with her daughter who had epilepsy. Mom would get up in the morning, especially in the summertime before anybody else and before daylight, build a fire in her cookstove. When she would get the fire up, she made biscuits, put the rolled oats on to cook our breakfast. And if we had any pork left over from killing the hogs the winter before, she'd fry that and make gravy. This was just a start for the day.

After getting through with the dishes and maybe making out some kind of bread to rise so she would have it for dinner (lunch), if she had any spare time at all—and I don't see how she had any—she would go out into the garden to pick peas, green beans, tomatoes, and things like that to improvise some kind of lunch for the farmhands.

If she got twenty minutes, she would run in to sew a little or hoe in her garden. Then she went back into that hot, little ol' kitchen area with a woodstove burning. After doing those things in the morning, she would repeat the process in the afternoon to get ready for supper. After preparing her supper and serving it, she would clear off the table, wash the dishes, and it would be bedtime by the time she got all that done. She would be lucky to get in bed by ten o'clock that night and I'm talking about Standard Time. And she was faced with getting up the next day, same time, and starting all over again with the same chores. This took place seven days a week. That gives you some idea about how rough her life was. But I never did hear her complain.

A loving, happy environment. Mom and dad also tried to provide entertainment for us. My dad could play the harmonica and the Jew's harp, as he called it. My mother could make the French harp talk, which I called a harmonica. And we would often sing after supper. The children knew that this was a necessary part of living. You know, in relating these experiences and thinking back over them as I have worked with you, my goodness, I could think of more and more stories and things to relate going into this. I could write a book about my life but John Steinbeck beat me to it with *Grapes of Wrath*.

GENEVA LOUISE BECKETT COTTON COLLINS

My best memories all revolved around my family. We were very poor, but a very close, loving, and large family. My mom had twelve brothers and sisters; my dad had ten brothers and sisters, all of whom had large families, and we were all very close. All my family were singers and most played some instrument. My dad played the fiddle, harmonica, and Jew's harp.

We had dances quite often. We had plenty of music and an uncle who called the square dances. I would sit back and watch my sisters, brothers, brothers-in-law, aunts, and uncles dance. We had the dances in our kitchen, with our tables, etc., pushed to the walls. I started dancing when I was ten years old.

Our house as the meeting place for everyone. Holidays, quilting bees, and any or no excuse to come. Quilting bees were a lot of fun. Women sat all around the quilt and who ever sat close to me always took time to teach me to quilt. There was always talk around the quilt; the women traded news of what was happening in their homes and with their children. My mom didn't gossip, so there never was gossip around the quilt!

ROBERT EUBANKS

As children, playing out in a yard that had been scraped clean with a hoe. We had no lawnmower. We would play at night after supper when it was a full moon way into the night. I also remember the good food—pork chops, homemade bacon, and ham. Occasionally when the cow would have a bull calf, we would have a little beef that we canned. No frozen food so Mother would put it in pressure cooker and can it. There's nothing on the market today that tastes as good as that home-canned meat.

Gigging fish. My dad liked to hunt and fish, and at an early age, I got to go hunting with him and run the boat for him while he gigged fish at night on the Tyronza River. There was also a drainage canal behind our farm that was also very fine for fishing. We would come in with all the fish that both of us could carry by eight or nine o'clock at night. On the front end of the boat we had a Coleman lantern and Dad had a big ol' carbide headlight on his head. He would probe the depths of the water. It had a bullseye lens over the flame. When that reflector was really clean and polished—which was my job—we would throw a beam of light a way out. Then we would work till midnight to clean those fish before we went to bed.

VIOLET HENSLEY

I remember the time we went to Meyer Creek where my aunt Alice had a Fourth of July picnic. We had watermelon and lemonade. Since we never had lemonade at home, that was a big treat. One watermelon from Dad's patch was so heavy—one hundred pounds—it took two men to put it into a wheelbarrow. I started playing the fiddle when I was about thirteen. We played for square dances every weekend from thirteen to eighteen and a half.

Why did you stop? For one thing, my husband had two left feet. But the real problem was that he didn't want me around any other man. He got over that later on when some of his friends told him, "You better let her work or she'll wither away."

GEORGIA HEARN HILL

I remember Grandma Hearn, my daddy's mother. She raised and sold turkeys. I don't remember how old I was. But Grandma had a white and red bandana around her head. She had on a long, white apron and she had me by the hand. We were going to follow the turkey to find out where the turkey's nest was down in the woods. Turkeys were very wise. They didn't want anybody fooling with their eggs. She would whisper, *"Be quiet! Be quiet!"* I remember that about my grandmother. I remember that when she died, Daddy came and got me. I went to sleep. And when I woke up we were back at the house. I was about five or six years old.

MARY FRANCES LOVELL IZARD

The day the peddler came. He was a marvelous person with a bus full of wonderful things. He did not have a schedule but one day he would arrive. We would run to the henhouse and get an egg, and he would trade it to us for a glorious piece of candy. My mother even liked to see the peddler come because in his magic bus he had things like needles and thread and other small things she had no way of getting. There is no doubt the peddler brought wonderful things to our lives.

KENNETH GUY LACY

There was nothing good about the Great Depression that one would want to remember, other than that we survived. I think the fact that people had to depend on and help each other made our generation one of the more patriotic eras than we have seen since. Because of enduring the hardships, it has instilled in us a strong fortitude of survival and equipped us with a greater sense of appreciation for the blessings we enjoy today.

EVELYN LANGLEY

When my mother had some extra money to get us a new dress, we's just tickled to death. A little twenty-cent dress would cost just ten cents a yard. Why, we was just as happy as larks. We thought we was really dressed up. And in the summertime, they had what'cha called "organdy material" back then. It was real thin. And let me tell you, it was real scratchy. My mother would dress us girls up in that organdy on all-day-singin' days, and we's just really dressed up, you know. She'd keep us in the shade all spring to keep us from gettin' blistered and tanned. You'd get up there for all-day singin' and run around in that churchyard and be blistered as red as beet pickles. Be sore as 'risin's for a day or two. But we just didn't think much about it. Mama would take the cream off the milk and rub it on our blistered arms and face and soothe it and make it not hurt so bad.

WILLIS MAGBY

The fall of 1936, when we were still in Miller County, we made twelve clear bales of cotton for sixty dollars, and we didn't owe anything of it. Dad could always brag about that. Right in the middle of the Depression—1936—he bought a team of mules and paid three hundred fifty dollars for them. After we got through with the cotton, he traded in two mules for a pair of young horses and got two hundred dollars to boot. He was happy but wasn't any happier than I was. He went into Texarkana and paid off more than five hundred dollars that he had borrowed for our groceries and for the feed we needed. He was the first person in Miller County that ever paid off his debts in a year. At sixteen, I was having to do stuff that most men couldn't do, but I had grown up with it.

DULAS L. MASSEY

In 1932 after I got out of high school my cousin and I went on a hobo trip to Oklahoma so we could pick cotton. We rode in an ice box on a freight train and got kicked off in Coffeyville, Kansas. We made it to the Hobo Jungle in Coffeyville for a bite to eat. They'd go around to restaurants and stores and they'd be given enough to make a big pot of beans, chicken, or meat that they'd been given.

I never knew of anybody ever getting hurt. There was no dope, anything like that. About every hobo in America knew where that was. That freight train was so full of people. People riding on top of them would look like a bunch of blackbirds on telephone lines. We had to walk all the way to Oklahoma City where my dad was working in an oil field. It took us about three days. For food, we ate peanut butter and crackers and we'd sleep on the side of the road.

One night we stayed in a police station and about got eat up by bedbugs. But we were only eighteen years old and didn't know any different. We each had a little sack on a stick over our shoulder and would walk along until somebody picked us up and we'd get in the car.

MAMIE H. MAYS

In about 1920 or 1921, we'd go into town twice a year—once in the spring and once in the fall. We would get a barrel of flour and a hundred-pound bag of sugar, things like that. We had our own pigs and calves for meat. People didn't go hungry. We had peach orchards and wild plums. While we were in town, we would always get something like stick candy. Every Saturday night in Lono, a man would come to the Overton Store and show us silent movies. Everybody would go and the man who run it made a lot of money selling ice cream. We'd come back after dark and we kids would lie in the truck bed and wonder why all the stars were following us.

End of WWII. I remember when Mama started shouting when somebody come by and told us that [World War II] was over. One of the reasons she was shouting was because it meant that my daddy would not have to go and fight in the war. They were fixin' to take the married men.

LYDIA MOSER RIDER

I remember in the 30's when the Depression came along, it was hard for everbody back then. It was terrible. I was 16, or 17 years old. Jim Twilley, Aunt Martha Twilley's son, and dad's cousin, worked up there in Melbourne in one of the garages where they sold cars. And he'd saved his money and he bought a brand new Chevrolet truck. It was a big truck, bigger than a pickup. He built benches all around the inside of that (inside the bed) and he put a tarp over it, just like a covered wagon. And all the people that want[ed] to go with him he brought out to Texas, and Arizona and New Mexico, and we picked cotton all the way through. And he took us clear to California.

That was the first time I ever left home. They was 30-some-odd that signed up and went in that truck. Jim would stop and see the foreman and get a job for all of us. We camped out mostly, or they furnished an old shack, or something that we lived in while we was picking cotton for these people.

We'd pick in one state till everything was picked up, then we went on. We carried our own sacks with us. We had a ball, but we worked and picked cotton all the way from Texas through New Mexico and then into Arizona.

[Author's note: Excerpt from "Down Memory Lane." Used with permission.]

MARIE TAYLOR

We didn't have much material things but did have a good family life. I can remember that on rainy days and on Sunday afternoons we sat around in the front room and Dad would sing folk songs and gospel songs such as "In the Garden," "Higher Ground," "Swing Low, Sweet Chariot," "Utah Carl," "Old Black Joe," "Carry Me Back to Old Virginy," "Swannee River," etc.

In the fall and winter, neighbors would come over in the evening; we'd pop corn, roast peanuts, and visit. Parents would play cards awhile and kids would play hide and seek, tag, etc., in the yard. Sometimes families would play checkers with a homemade board and buttons, dominoes, and later Chinese checkers.

EVA SMOKE WELLS

When Roosevelt became president. For the first time in years, there was hope. Everybody kind of picked up and brightened up a bit. There was hope that he could do something rather than just talk like Hoover did. I remember the WPA and a number of other organizations that started to help farmers. They would pay someone at the county level to come out and build terraces to stop our ground from washing off. I've still got one right down here [pointing] that was built back in those days. It's still there. It never eroded.

They did a lot of good that way. They started programs where if you wanted to plant an acre or two of cucumbers and sell them to the pickle factories, they would furnish the seed. The government always made sure that there was some place that would buy what we produced. There was never any guesswork about it. They had contracts that these places would buy that stuff.

The way we stuck together as a family and shared whatever we had. I learned a lot that helped later in life. I remember when Dad went down on the river and saw a big fish trapped in a hole of water. He caught it, brought it home, and we had a big fish fry.

My best memories also include my parents, who worked hard all their lives. Mom was eighty-four when she died in 1990, and my dad was eighty-eight, when he died in 1992. He died on my birthday. They left six children who not only know how to work but we really enjoy it. And my four children were taught that a good day at work brings a lot of satisfaction.

ALBERT M. WILLIAMS

The hard times in retrospect. After looking back over my experience during the Great Depression—the closeness of my family was a good thing. I wouldn't trade it for nothing in the world.

FLOYE WINGFIELD

How I would look down the road and see somebody coming. Your eyes would get bigger and bigger wondering who it was. And the joy there would be in finding out who it was—a member of the family.

What kept your family going? Each other. There was always one of us who needed something from somebody else. And no matter how small, it was never overlooked. It was just automatic that you were going to do it. Even in the fifth generation, my family is still like that.

Chapter 29

Funniest Memories

GENEVA LOUISE BECKETT COTTON COLLINS

Drunk chickens. One year, my youngest brother, Dub, and my brother-in-law, Sammie, dipped the froth off the cooking cane juice. Daddy told them to dip off the top and get rid of it. They took the froth and instead of getting rid of it, put it aside. Time passed—I don't know how much—but Dub and Sammie started dipping into their ill-gotten froth and became very happy. When Daddy realized what they were doing, he ordered them to get rid of their "new drink." They did, and later my mom called Dad to come and see her chickens. They were staggering and falling down. Daddy asked Dub and Sammie what they had done with the dippings and they told Daddy that they had put it in Mama's chicken feeder. So Mama's chickens were drunk!

GENEVA KING EMERSON

Things were still awfully tough when WWII began and we were hit by rationing. Elastic to put in the bands of under garments could not be had. Buttons and ties were plain unhandy. But if someone in the community could obtain an old tire inner tube that was beyond patching, the ladies were in business! Through trading, or somehow, my grandma got hold of a piece of inner tube which she cut into strips and used in the bands of our bloomers.

My aunt was of courting age. She and her boyfriend were standing at the end of the house next to the cistern, as far as they were allowed to go from Grandma's sight, for privacy. Suddenly, my aunt felt the innertube strip break and her undies started falling down her legs. Quickly she pointed to something over by the garden, diverting her beau's attention, stepped out of the dropped garment, and kicked it behind the cistern. Hey, we learned to be innovative and quick on our feet in those days!

VIOLET HENSLEY

I've killed lots of rattlesnakes and copperheads. I carried a live king snake around my waist for a belt. They kill copperheads, you know. They swallow 'em down to smother them to death. I'd started after the mule one mornin' to plow and I heard this rustlin' down in the grass and kicked it and I watched until that king snake had spitted the copperhead out dead.

GEORGE MERTENS

Turning the chickens. I remember hearing a story about a boy's chore before he went to bed. It seems this particular farmhouse had a breezeway down the center that contained a flour barrel. The chickens liked to roost on the flour barrel so it was this boy's job to turn the chickens if needed.

Black-snake scarf. I was tagging along with him and my dad, who were hauling loose hay. Dad was on the ground, pitching it up to my uncle, who was on the wagon. Dad speared a black snake with his pitchfork and asked my uncle how he would like the snake around his neck. My uncle said that would be okay, not thinking there was a chance that would happen. But when Dad threw the snake, it wrapped perfectly around my uncle's neck. Well, Dad and I thought it was funny!

WILLIE MORRIS

Why "Goosey" jumped off the railroad trestle. The railroad track ran over the Bayou Bartholomew. Across from the railway on the bridge was a place for the wagons and the cars to go across. We boys a lot of times liked to walk across the railroad tracks. Back in them days, one of these boys was what we called "goosey." When one of the other boys sort of "goosed" him in the behind, he jumped off the trestle and fell fifteen to twenty feet down into the bayou. It didn't hurt him.

UNION H. STOUDAMIRE

We had a little horse that got out. Me and my husband chased him around until three o'clock that evening. That horse ran me and J. C. all day trying to get him back in the lot. When he got tired of running he would walk for a while. Finally he got to where he seemed to stop and say, "Y'all tired? I am, too!"

EVA SMOKE WELLS

The black snake. My dad was plowing corn one day on my great-grandfather's place on Highway 51. A big black snake came from in front of him, ran up the leg of his overalls, out the other side, and kept on going up the row of corn.

ARWILDA WHITESIDE

We or Mama were not allowed to wear lipstick. Mama would order makeup from "Sweet Georgia Brown" through the mail. She would hide it from Papa in the quilting bag. One Sunday, I was nine or ten years old, and I saw a purse with makeup in it. I hid under a pew at church and made up my face. I sat up! Papa saw me and took me to a water pump outside and whipped me good and washed my face in cold water. I must have looked like a clown. I still hate to wear makeup!

CHARLES WHITFIELD

Everybody wanted their cows dehorned and pigs castrated, but they didn't want to do the work. I went down to my brother-in-law's place. I think it was fifteen cows that I dehorned that morning. I never did get a penny for none of it. And I was all the time castrating their hogs for somebody. Pluto lived over close to Mrs. Hodges. He had a lot of hogs, about ten or twelve sows. He said to me one day, "Charlie, would you mind coming over and castrating a pig or two for me?" I said, "Yeah, I can do that fer ye." I got over there and he had fifty! I castrated pigs until twelve o'clock. And you know, he never thanked me for it, let alone paid me fer it?

I remember that one time that O'Neal Cook had a big ol' Hereford male. He had big, bad horns. When O'Neal would feed his cows hay, this ol' bull was peaceable with the cows. But when O'Neal would feed 'em, that bull would hook them and cut up them cows with his horns. O'Neal asked me what I would do about it and I said I would cut the bull's horns off. He said, "Cut his horns off? How would you do that?" and I said, "Just throw him down and cut his horns off." He said, "Huh! You'll never throw that old bull down." I said, "Why I can throw him down by myself!"

So he told me to come on over and cut them horns off. I went over there and the old bull was gentle. We caught him, tied him up to the barn. I told O'Neal to run his tractor up behind him where I could tie his hind legs. I tied a rope around his neck and forelegs. You're supposed to just pull it and cut off circulation in his legs and the bull will just lay over. I got to pulling on that rope and the old bull just laid down. I tied him and cut his horns off. O'Neal never figured out how I could do that.

LUCILLE RIDER WILSON

A demonstration that went wrong. I remember we were in a Sears store in Fort Smith to buy an electric wringer-type washing machine. To show us how safe it was, the salesman put his Panama hat into the wringer and it sprung the wringer open. When we got home and set the new washer on the front porch, my brother Coy wanted to demonstrate this safety feature to our neighbors by throwing in his summer straw hat. It was crushed completely as it went through the wringer. When I think of this, I laugh hysterically because it is so sad!

FLOYE WINGFIELD

The problem of holding a goose. In the fall of the year, Mama would pick the geese to make her beds and to plump up her pillow. You'd wash them, clean them, put the feathers in a sack, and save them until you needed them. There was a bench with a dishpan under it beside of the house that she sat on to pick the goose. She would get that old thing and put it on her lap, turn that belly right up to her and stick its head under her arm. I was to sit on the other side and hold that stupid goose's head all afternoon.

Well, I wondered why I was having to do that while the rest [of my brothers and sisters] was playing? That went on for a long while until my little hands got tired. "Well," I said to myself, "it's time to find out what that goose will do if I let go of its head." So I turned its head loose. "Ooops! I'll never do that again!" [laughing] It twisted a plug out of my mama's arm. She came back with that ol' goose, done her up good, and then she got me. I tell you, when she finished with me, I woulda held the feet, too, if she wanted me to. I knew right then that I wasn't ever going to be a goose holder for the rest of my life! I was going to do something else to make a living.

Chapter 30

Other Memories

CELIA ALMYRA GRAHAM ACREY

How Mom spent her fifty-six dollars inheritance. The children of Dan and Addie Smith [Mrs. Acrey's parents] were left a small inheritance by Addie's grandparents, Jessie and Mary Smith. Apparently, they left Addie five hundred dollars, which was passed down to her children, since she was already deceased. The five hundred dollars was divided equally between the five living children. The oldest children, Nathan and Luther, received their money at that time; however, Celia, Edith, and Delton's portion was put in a bank in Searcy, until they reached the age of eighteen. Before they turned eighteen, however, the bank went "bust" in the fall of 1932. Apparently, however, they each did receive a portion of the money.

Mom received fifty-six dollars, which she will tell you she "blew" on the following: a black coat with a fur collar from Sears Roebuck; a white kitchen cabinet (fifteen dollars) from Robin Sanford's General Store in Searcy; a bedstead, mattress, and springs.

[Author's note: As told by Linda Acrey Ashford, Mrs. Acrey's daughter.]

KATHRYN BAJOREK

Cucumber babies. My mother saved old yellow cucumbers. Mixing flour and water, she made a paste and glued cornsilk in different colors and we had cucumber babies—blonde for a mama doll and brown for a daddy doll. In the spring, my mother would plant cucumbers and let them grow until they were big enough to make the "babies." I never had a doll from the store. My cucumber babies were special. When they got soft, they would die and my mother would wrap them up in cloth scrap and we would bury them in the garden.

Getting us ready for church. My mother would French braid our hair on our way to church. It took us forty-five minutes to an hour to get there, depending on the mud. We rode an iron-wheel wagon pulled by two mules. Mother always took wet rags to make sure our faces were clean.

It was rough before we got in the wagon to go to church. The cows had to be milked and water pumped for the cows and hogs. We never had a clothesline. We hung clothes on barbed-wire fences. We made our own lye soap, so that is what we [used to] wash our hair and bodies with. To have a good life, my advice is to take life as it comes. Everything happens for a reason. God has a master plan for our lives. In all things give thanks.

FRED NORMAN BLANKENSHIP

Ghost stories and Aesop's Fables. My grandmother took the weekly *Kansas City Star* that came every Wednesday. There were two things that were always in the *Star*—a ghost story and Aesop's Fables. That night I would go to her house. Although I could read, I liked for grandmother to read the ghost stories to me because she could make them sound scary. I would have been disappointed if neither of those made the paper.

My first steak. I remember one time when I stayed all night at my grandmother's house. The next morning she fried me a beef steak after my daddy had killed one of the cows. That was the first steak I ever had and it was good.

Daddy's birthday. One other day that was important—May 26, which was my Daddy's birthday. Two weeks before his birthday, Mother would put a hen in a coop and really feed the hen for two weeks. By the time we killed her, she was really fat. On his birthday, my daddy would make it a point to have English peas and new potatoes. She made the best chicken and dressing I ever ate.

Two family traditions back then. One was at Easter time. We'd go to Pilot Knob Mountain near Melbourne. We'd go to the foot of it, eat, and then climb the mountain. It was hard to do but exciting. We could eat all the eggs we wanted—fried, boiled, or any other way you wanted to fix 'em at home. During all the the other weeks, we'd gather the eggs and take them to the grocery store and exchange them for sugar, coffee, and tobacco. The other place we went was the big Standing Rock formation on the upper Piney Creek in north Izard County.

LOUISE J. BRANT

Watermelon hot caps. My dad, by using hot caps, always tried to have some melons ready for market for July Fourth at a cent per pound. After the price dropped so low, we kids were permitted to sell what we could. Once I "hit the jackpot" and loaded a school bus from St. Paul, Arkansas, with fifty large ones for two cents each. A whole dollar at one fell swoop!

R. L. "BILL" CARTER

Poke and Grits. We'd have for breakfast gravy made out of water and flour and salt. We called it "We Like It." [chuckling] We HAD to like it! I had great uncle Joe Carter, he called it "Poke and Grits." He said he had to poke his feet under the table and grit his teeth in order to eat it. [laughter]

Telephone service. They had a telephone line going from Monroe to Brinkley. The telephone operator was Minnie Blue. We didn't have a phone in our house but there were two phones in the community. A lot of times, Minnie Blue would be running over to Monroe and we couldn't call anybody. Anyway, the old man that owned the phone company occasionally he would walk the lines from Monroe into Brinkley, which was about fifteen miles. He stopped by our house one day. I was a little old kid sitting on the steps barefooted. He reached and grabbed my foot and started biting his thumb and I thought he was biting my toe. [laughing] I never will forget that.

Peddlers coming through. They was two or three of them who would come through once in a while. One of them would be selling the *Progressive Farmer* magazine. We'd help him run down an old hen for the payment. One guy come through and he was selling vanilla flavoring and stuff like that. After we moved up to the eighty-acre place, a guy came by in some kind of vehicle. He was selling dry goods. He'd take anything you had except money—we didn't have no money. [chuckling again]

We sure did hunt—ducks, rabbits, 'possum, 'coon, deer. We trailed through plum thickets and mulberry trees and blackberry patches and wild onions. I've heard people say they wouldn't ever eat a 'possum but they've never been hungry! I've eat a lot of 'possums. We go down the road now and see one that someone has run over and I tell my wife, "Now there goes a good meal." [more laughter]

How did R. L. Carter get the name "Bill"? When I was just big enough to crawl and I fell off the bed and hit head first in a bucket of water. Dad jumped up and grabbed me and said, "Get up from that water bucket, Bill!" And they called me "Bill" ever since! *Why?* Don't know.

DOYLE L. COLLINS

Family decides to go to California. Because of continued crop failures due to droughts and the poverty that was created in Arkansas, and considering the money that the older son Elijah was making in California in the winter of 1938, the family discussed selling out and going out there. By pooling their income, they felt they would make more than they were making on the farm.

Even though they were discouraged by other people who had tried but failed to find their fortune in California, the family finally decided to do it. Being ten years old, I was gleeful they had made that decision because for me, it was an adventure. We sold all the farm implements and the livestock and even some of the furniture and bought a 1936 Dodge pickup truck. My dad improvised a wooden, camper-type structure that had wood benches on the back of that pickup with a window on each side and a storage under the benches for us kids to sit on during that trip. Looking at the trip today, I wonder how we made it. Can you imagine all that we would need for hauling all of the family, nine of us—three girls, four boys, and our parents—in the back and in the cab of that truck?

We left out before daylight on Easter Sunday 1939. My mother and dad were in the front with my older brother and the rest of the kids were in the back. If we needed to go the bathroom, we would pound or tap the cab.

My mother, of course, had taken quite a bit of cooked food with us. We would stop and eat on the side of the road or eat in the back of the truck or in the cab as we went along. We also became acquainted with baloney and crackers.

During our journey, I had my first exerience with motor courts (what we call "motels" today). If I remember correctly, we paid three dollars for lodging for the whole family that night in Groom, Texas. It was also only the second time we had run into flush toilets and running water. We'd already used them in service stations.

We drove on up into the mountains and around those hairpin curves in western Arizona. The ascents and descents were really scary looking to us, even though we had been raised in the hills. The next day, after getting up early, we ran into a snowstorm as we crossed the Mojave Desert. After driving out of the snow, we stopped at a pull-off and looked at the valley below us and how green it was! We were also able to see the town of Arvin six or so miles further, which was our destination. In *The Grapes of Wrath,* the Joads pulled out at the same spot! When we arrived into town, Dad and my older brother found jobs pruning and thinning grapes. My brother, who was two years older, and I went to school.

That summer, we saw every type of conveyance, including trucks and automobiles, and viewed the improvisations that took place with those dream-seekers from Arkansas, Texas, and Oklahoma. Their tents, automobiles, lean-to's, whatever, were parked on vacant lots. Families lived in those tents in the summertime when temperaures of 110 to 115 degrees were common.

Family returns to Arkansas. Because things had not worked out as we had hoped for, we came back home on November 1, 1939. The older members of the family—the girls in particular—were happy to get back to their friends, but since I had become acquainted with soda pop, ice cream, indoor toilets, and being able to take a bath and a shower occsionally, I didn't want to leave those simple pleasures. I felt a sense of failure. We had gone to California to seek a better life but we had failed in our mission. I'm sure my parents felt the same way, probably more deeply than I did, but they never conveyed that to me. You can imagine how hard that must have been for them.

GENEVA LOUISE BECKETT COTTON COLLINS

A home full of love. We had fun popping popcorn in the fireplace, baking sweet potatoes in the ashes, listening to my mom reading out loud from what books she had. She would take us away from our farm in our imaginations to far places and with strange people; she loved reading, which I inherited from her. She would boil sorghum molasses into candy and we kids pulled [it] into taffy with buttered hands. I had (I believe) a wonderful life. You can't beat a home full of love.

WILLIAM EDWARD DELAMAR

How I learned the value of money. I used to go out and help my daddy when he would get a job. I was at home. I didn't pay no bills. I just worked with Daddy. I would see the money they paid Daddy a lot of times but I didn't know that Daddy had to pay bills. So one day I asked him for a dollar. Daddy was a quiet, easygoing man.

"I don't know, son," he said. "Let me look and see." He gave me a dollar.

"Papa, you know what? When I get grown and get on my own and make money like you do, I'm going to have some."

"Come here, son. Sit down," he replied. "Listen to me, now. Listen to me. When you get grown and get on your own and have one dollar in your pocket and don't owe nobody nothing, you're doing all right. I don't owe nobody today. I haven't bought on credit for twenty-six years."

I never owed a bill in my life I didn't pay. Since I have been here in Malvern—I came here when I was twenty-one or twenty-two years old—with that ten dollars I borrowed as a teenager, and the twenty-five dollars I borrowed in Malvern, that's a total of thirty-five dollars. That's all the money I have ever borrowed.

There's always been somebody worse off than you. Back then, people who didn't have anything would get the most help. I know that back in WPA days, when people were working on the roads, a lot of people got jobs. But my daddy couldn't because if you had any way of helping yourself, you couldn't get a job. To get a job, you would have to have nothing at all.

I see people today who are on welfare, but we've never been on welfare and we've never had food stamps since I have been grown. If I was lazy and wanted to cheat, then I could lie and make people think I was poor. Let's put it this way: People who are not handicapped will sometimes park in handicapped places. I don't do things like that, either. I don't do no cheating. I try to treat everybody fair.

What gave you the courage and the strength to survive the Great Depression? It was my mother and father. I was just a kid. But when my mama and daddy got down on the floor and prayed, the building would shake. There was something about my mother's prayer that has kept me here today. I listened and paid attention. Most of the time a disobedient child will wind up in trouble. Nobody can tell 'em nothing. But every time I got up to go somewhere, my mama would school me before I would leave. "Son," she would say, "don't you do this and don't you do that." I don't know how she knew what I had in my mind but it was something that I had been planning on all week. "Listen to Mama, now," she would say, and I would. But I was blocked from doing the wrong things that I had wanted to do!

I have goodwill towards everybody. I don't create no kind of hate towards anybody. I'm not too bad about hating anyone. One thing that makes me be that a way is that I trust in the good Lord and I know God is good to everybody. He's good to me and He's good to you. He's good to all people but some don't accept Him.

God makes no difference between you and me. We can be friends. If we cannot be friends, it is because we do not want to. I can help you and you can help me. We can't live hating colors. There are all kinds of colors in this world. Our clothes are not the same color. We don't or shouldn't base ourselves on color, but on who we are and what we stand for. I just try to do it right and love everybody.

S. LORETTA ECHOLS

The WPA project built many roads. Before it came along, for example, you couldn't get from Stuttgart to Brinkley because of the swampy areas. But a road was built by the WPA. Some of the bridges were two and three miles long and are still being used today. My dad and uncles worked through the CCC camps and WPA on community projects. If it had not been for some of those New Deal programs, families would have suffered much more than they did. The government programs in the New Deal and the First One Hundred days were lifesavers for the rural areas during the Depression.

JEAN EDWARDS

We had sixteen tenant families on our place. They would work the land with mules. When the war broke out in '41, my father had to make a decision on whether he wanted to continue to farm or do something else. He had invested in land, so he decided to drive a tractor. He bought a couple of tractors. He did that because they had opened up the Pine Bluff Arsenal. They were paying seventy-five cents a day on the farm for labor. The Arsenal opened up paying $1.25 an hour. I and a friend named Curly became the tractor drivers at the farm. At that time in '41, I was twelve. Curly and I became Daddy's chief workers.

This meant that I had to drop out of school. I wanted to work out some way to work out cooperation with the school to accept some type of educational plan. And Mr. Harris agreed that if I could make my English classes, he would figure out a way that I could turn in class assignments and reports and get credit for the rest of the work and my education would continue.

As a result, I would go to school every morning and my first class was English. Then I would go back home and go to work on the tractor. I did that for about three years. Because of that, I was able to obtain my high school diploma. Mr. Harris knew how much I appreciated what he had done for me.

> [Author's note: Mr. Edwards not only survived the Great Depression but in 1990 became only the second African American to be elected to the Arkansas Senate since the Civil War.]

ROBERT EUBANKS

Cracklin' and "fatty" corn bread. Fatty corn bread was made with cracklins and was real crusty on the outside. It wasn't baked in a regular bread pan but in a cast-iron skillet in the oven. Mama would make out a patty and she would pat it down in the skillet with her fingers, so it always had finger marks in it. This was the richest corn bread you ever ate in your life. Those ground-up cracklins in that corn bread made for a very rich piece of corn bread. You could go out on a hunting or fishing trip and put a pone of that corn bread into your pocket and a turnip and brother, you were set!

ROBERT FLANNIGAN SR.

Some people blamed President Herbert Hoover for the Depression. Well, the

Depression came on and he just happened to be our president when it climaxed. It wasn't his fault anymore than it was that of other people. It was just the times in the world. And when Franklin Roosevelt went into office in 1933, the government began the New Deal programs to put people to work, such as the WPA and the CCC. But the government offered people jobs, and they built the pavilion over at Walcot and the bridge down in south Arkansas across a slough.

MARY EATHEL FREEMAN

A lot of people weren't able to pay the taxes on their land, which went back to the state for delinquent taxes. It could be given to anyone who needed a home. You first had to find acreage of state land, and then sign up by agreeing to live on the land and make improvements by clearing and building structures ("proving up"). After a period of time, you would then receive the deed for a small fee. The term for this process became known as "donating."

In 1933, Aunt Janie and my grandparents were living on the Ross place, one hundred and twenty acres of land "donated" by Grandpa Ingram. Neighbors helped build a house. There was already a barn but they added a chicken house. When the time came for proving up on it, he was unable to pay the small fee, so Aunt Janie signed up to donate it. She managed to pay the fee and received the deed, so the land was hers.

[Author's note: Excerpt from Eathel Freeman, *Journey of a Depression Kid,* 3rd printing (self-published, 2005), 16. Used with permission.]

Skinny dipping at night. Nellie, my sister-in-law, got by with things my sister Lela and I were not allowed to do. One time, we wanted to go swimming. My mother said, "No." She thought playing in the water was unhealthy. Later in the evening, Nellie, Lela and I went to drive the milk cow home. We decided to go "skinny-dipping," although at that time, we had never heard that term. We jumped off the rocky ledge, climbed out, and jumped in again. It was marvelous fun! Then we heard a dog barking. It had treed some animal. Three females were dressed in record time and hurrying home with the milk cow.

[Author's note: Excerpt from Eathel Freeman, *Journey of a Depression Kid,* 3rd printing (self-published, 2005), 23. Used with permission.]

My husband spent some early years in the Civil Conservation Corps. It was an experience that helped him all through life. Arden and the crew set out bermuda grass on farms to get it out of cotton which did not hold the topsoil. He and crews built terraces, set out pine trees, vaccinated cattle, I think. I know he did when we had beef cattle. He learned quite a lot about soil conservation.

The moonshine industry. Making moonshine was prevalent in the foothills of the Ozark Mountains. My friend Heather Oliver said it was the way most people fed their families. My dad, Ben Campbell, never made any. So we were the poorest people in the area. I would have liked to see [a still] at work but with my luck the "revenuers" would have come on that day—Ha! I've never been able to get by with much. Maybe because I don't try much. Arden said the revenuers usually caught the makers and took them away from their families. In 1965, when we had so many hospital bills and two children in college, I almost wished we could make it!

One of my best friend's husband made good whiskey. He sent five children through college. Another friend said her father made [moonshine] and that it paid for food and clothes. When he was caught, they sent him to prison [but] he left enough booze to take care of her and the children! Ha!

ZELMA LOUISE HAMILTON

We had to walk two and a half miles down a dirt road to the school. One day after a big rain and several wagons and a few cars traveled the road, deep ruts were made by the wheels of the wagons and cars. The ruts were soon dried and very hard. My friend and I got to quarreling walking home from school. I grabbed her by the hair and slung her on the hard ground. Of course, I was walking ahead bragging about what I had done and not paying attention to her. She crept up behind me and shoved me face down on those hard ruts. We both were crying. Before we got to my house, we were "arm in arm" best buddies. We knew that we would get a whipping if our parents found out about the little squabble.

Boy stricken by incurable hydrophobia. After a boy in our school had been bitten by a rabid dog, for some reason, he didn't take all the shots. One day he ran up behind my sister, grabbed and kissed her. Three days later, he was diagnosed with rabies. Hydrophobia was incurable at that time, so after the boy's parents had given their consent, two doctors each gave him a shot, one of which was to take him out of his misery. Neither of the doctors ever knew which one had given the boy the shot that killed him. We watched my sister for a month, not realizing that he would have had to bite her to get the germ into her blood stream. I don't know if that could really happen.

ZELMA LOUISE HAMILTON

But I still had to pick cotton! One day my mother was baking cookies. My brother, sister, and I were waiting for our first cookie. After I had eaten it, my mother told me to go pump my bucket of water. I didn't want to, so I went to the pump, jerked up and down on the handle to prime it. My hands were greasy from the cookie. The handle slipped out of my hand and hit me under the chin. I was holding my wrist and swinging it at the same time. My dad yelled, "She broke her arm, she broke her arm!" I hadn't but I sure had a sore chin. I thought this would keep me out of the cotton patch. It didn't. Mother wrapped a rag around my head and chin and I picked cotton. I was about eight years old.

VIOLET HENSLEY

How to make corn-shuck mattresses. The pillows we made were either from picked-off cotton stuffed in a bag, or chicken feathers. We would usually save the fine chicken feathers since we didn't raise geese. But for mattresses, back when I was big enough to pull shucks apart—we'd pull them into strips, because you know a corn shuck has this big ol' knob at the end that joins the stalk. You tear that out because that would be like sleeping on a rock. We also pulled grass and stuffed that inside the mattresses—we called them bed ticks—anything to fill them up. We had iron bedsteads with boards set across it to lay the mattress on. Later on we made a cotton mattress with some bed springs.

[Author's note: In the midst of the Depression, Violet began to make and play fiddles (see "Earliest Memories of the Great Depression"). At the time of the interview, she had completed about seventy-one, she said, which sell for as much as three thousand dollars each. Appearing each year in Silver Dollar City in Branson, Missouri, she has worked with such artists as Glen Campbell, Jimmy Driftwood, Senator Robert Byrd, Stacy Davenport, and Ricky Skaggs, and sold a fiddle—soft maple back and basswood front—to Senator Byrd, who called it his "favorite" and used the fiddle during a performance at the Grand Ol' Opry. In 2010, during a tribute to the late Bill Monroe in Owensboro, Kentucky, Violet was named as "One of the legends of bluegrass music."]

MARY FRANCES LOVELL IZARD

When people are hungry they don't kill wildlife for sport but to eat. There were a few rabbits and squirrels but the deer and bear population were wiped out. Some people ate 'possum and raccoon but my mama refused to cook it. She said it wasn't clean meat. We were visiting some friends and we were invited to stay for supper. We all gathered around the table and began passing the food around. There was one big bowl of squirrel and dumplings and several other bowls of vegetables. Squirrel and dumplings were one of my favorite dishes. When the bowl got to me I spooned some in my plate but when I looked down I saw they were floating with hair from the squirrel. I lost my appetite. That was when I learned that everyone didn't clean and cook their food like my mama did.

ROSCOE E. JEFFERSON

Driving hogs from Yellville to Pyatt. We had a feller come in and bought hogs. We lived twelve to fifteen miles from Summit to Pyatt. My brothers drove them hogs to Pyatt and loaded them on the train. The cows didn't bring enough money at the market to pay the freight bill. "Well, all we can do is to send you another load of hogs," but the man didn't want any more. That's how hard it was. So I guess the bill went unpaid! [chuckling] Sometimes we worked for other people or sold eggs or traded.

Wheat harvest. I was probably in my thirties. I went to the wheat harvest in Colorado when my son Truman was a baby to work in the wheat fields. The government sent us out there on a train. It was back in the time of the war (WWII). The government needed that wheat to feed the soldiers. We slept in the barn. The only time we went into the house was to eat meals. I believe we were paid seventy-five cents an hour. They'd pay our way out there for thirty days and pay our way back.

I worked on building the railroad when I was twenty-one. We were paid fifteen cents an hour. Another way we had of making money was to dig mussel shells out of the river and sell them. We'd get a little money. It wasn't much.

KENNETH GUY LACY

The New Deal and "snake-head" shoes. One incident that I remember concerning new shoes was after President Franklin D. Roosevelt began to pull the nation out of the Depression. As part of his New Deal program, centers were set up in communities that included canning kitchens and mattress making. I remember my mother went to the closest one (about four to five miles) and got me a pair of shoes called "snake-head" shoes. They were like a work shoe, having a piece of snake-skin-like pattern across the toes. I don't remember if there was only one pair of shoes per family, but I was the only one that got shoes at that time. Maybe I was the only one that really needed them, but I surely was proud of those new shoes.

New Deal canning kitchens. Families from miles around would gather their fruits and/or vegetables, maybe meat, and take them to the center for processing. The people in charge had pressure cookers in which they canned the produce that was brought in. Woodstoves were used to heat the cookers. I remember something about metal cans being used in the beginning. I also remember Mom taking glass jars to be used. Sometime during or maybe after the program ended, they gave a cooker to some of the families, or maybe all famiies got one. Mom got one, which she continued to use as long as she was able to can. It could have been the cooker was given at the same time we got a horse, harness, and a plow. The cooker is still around.

New Deal nattress making. The government provided sewing machines, cotton, dark-blue stripped mattress ticking, and needles that were long enough to go completely through the mattress and instructors to help in the making. I don't remember how the number of mattresses a family could make was determined. Perhaps we got one for each bed—four. We have a re-conditioned mattress that is still in use that has cotton in it that was used in a mattress made back then. Springs were put in it, making an inner-spring mattress.

Each family was given an animal or two, and, depending on their need, a harness and a "Blue Bird" plow. I have the plow in my barn. Some families got a pair of horses or mules if they had none. We had a mule, so we only got one—a mare. They didn't make a good-looking matched team, but they worked. The animals were shipped in from out west. They were wild and had to be broken. Our mare wasn't so wild and was easily broken, but she would "sull" when she didn't want to work. We named her "Pet" because that's what she became. She ended up making a good working mare.

SUZANNE GROSS MARKS

My brother was called "Tarzan." He had a nice singing voice. We had a mature night at the school gym. I made a pigskin costume for "Tarzan." His act was to make bird calls and holler like Tarzan. The audience liked it so much they wouldn't let him go. So he said, "If Miss Mable would play the piano, I will sing 'Mother.'" Tarzan sang: **M** is for the million smiles she gave me, **O** is for the other things she does. **T** is for the tears she shed to save me. **H** is for her heart of purest gold. **E** is for her eyes, with love-light shining, **R** means right, and right she'll always be, Put them all together, they spell "MOTHER," A word that means the world to me." It brought down the house!

DULAS L. MASSEY

CCC camp experience in 1933 and 1934: I was one of the first ones that went in. We got thirty dollars a month. We had to send twenty-five dollars of it home and we got the five dollars that was left. We built roads, fire towers, worked in the forest.

How I stopped my dad from bootlegging. In 1929 when I was about fifteen in Scotland, which is the hills of the Ozarks, this was the seat of all the whiskey and bootleggers. A lot of people, including my dad, had some interest in bootlegging at one time to make money. I was the cause of stopping his bootlegging I guess. He had eleven and a half gallon jars of bootleg whiskey under the barn. I didn't know it was under there until a friend of mine was crawling around under there and there it was.

I didn't drink any of it but my friend did, probably didn't think a lot about it, and later on went to town with some bigger boys to where it was. We was swimming there in the swimming hole about twenty feet from the barn. We were both as naked as we could be when we were crawling under the barn. Later on he went into town and told the big boys and they come up there and stole ever bit of it. Most of the boys [who] were in high school got drunk that night. It liked to kill ol' Floyd Blue. Daddy was going to give me a big whippin'.

He usually didn't whip me very much but he really got after me there and pushed me around. We had a lady there that stayed with our mother. That was the day our sister was born. The lady told my dad, "You leave that kid alone! If you weren't foolin' with that stuff, he wouldn't have got it!" He was so embarrassed that he didn't lay a hand on me!" [laughing]

MAMIE H. MAYS

All the people who went to California and come back had money and a car. Even though my daddy had no idea about how far it was to California, he decided that we were going there and we did in 1925. He said it never rained in California, that they watered everything by irrigation.

He bought a new Ford truck that didn't even have a cab on it but had a windshield. There wasn't a seat but just a cushion to sit on on top of the gas tank and that was for the driver. And he had a big, wide bed on the back of the truck. He put two cotton mattresses that we had and packed our dishes in a barrel and off we went—Daddy, my mother, my older brother, my younger brother, my sister, and me. He bought a little kerosene cookstove and we stopped along the highway to eat. That's the way we went to California!

When we got to Texas, there weren't any motels. Everybody camped on a campground. A man who was traveling in a coupe told my daddy, "Mr.

Heard, you better let your wife and those two younger children ride with me because when you're going across the desert, it gets awful hot." Since there wasn't a top to the truck, we went with him and that helped us a lot.

My brother was two years older than me, and he had learned to drive a car. The tops of his and Daddy's hands blistered while they were driving the truck across the desert because they didn't have any top! My mother and me and my younger brother rode with this other fella. We'd go along and kind of keep in touch with them. And that's the way we got to California. Everybody did! We camped on the side of the road.

When we got to Orange, California, we had friends there. Mama got a job peeling apricots. My daddy got a job pruning grapevines and all that sort of stuff. Me and an older cousin from California kept the kids. I was baby-sitting but didn't know what I was doing. My cousin in California was the meanest little girl I ever saw. They called us "Arkies" and "Okies." We didn't know they were making fun of us!

In California, we lived in tents, and my brother and I had to walk about three miles to school. Daddy got a job working on steam engines to pump the oil wells. But we came back to Arkansas in 1926. My mother was pregnant at the age of forty-four and was afraid she couldn't get the right treatment in California.

MARION ANDERSON McANALLY

My father and mother came from poor farming families and neither of them had very much growing up. My dad grew up near Fendley where he farmed about one hundred fifty acres and grew primarily cotton and corn. My mother, who was the daughter of a Confederate soldier and a very young half-Cherokee woman, was born when her father was sixty-six years old. Life was very difficult on the land that my grandfather had bought around the turn of the century for fourteen cents an acre. Mom put it like this: "We didn't have anything before the Depression so we hardly knew there was one."

Hard times. Dad and Mom married on September 9, 1933, in the back yard of a home at Fendley. A few days later, they set out in an old Whippet automobile for Missouri to pick cotton. The old car had huge wheels with wooden spokes which rattled as they drove. They had to periodically stop and soak the wheels in a creek to make the spokes swell so they wouldn't rattle.

As they neared the Arkansas-Missouri line they had a flat, one of many they had along the way. The old tube had been patched until it couldn't be patched again. My dad pulled the tire off and stuffed it with cotton boles from a nearby field. This makeshift tire got them to their destination in Haiti, Missouri, where they spent their honeymoon—picking cotton—while staying with a couple from Alpine.

Keep your one hundred dollars! A man in the church had an affair with a woman in the community. He had a habit of giving his one-hundred-dollar offering once a year and making a big adoo by standing and counting it out in one-dollar bills. In 1929 his offering was publicly rejected by the church because of his behavior. That was a lot of money for the church to refuse when it had so much debt. The church, however, was blessed and the building stood until 2006 when it was destroyed by fire."

[Author's note: Information on his mother and father was provided by John McAnally, Bismarck, Arkansas.]

RUPERT E. "BUSTER" MELTON

My dad owned a bunch of land [and] had plenty of money. He loaned it to neighbors who just didn't have no money. If they wanted any money—$100 or $200—my dad would loan it to 'em. They needed it to buy groceries, stuff like that. Of course they didn't have to have all the groceries like they do now. They had stuff to eat like fatback, canned stuff. We didn't know what store-bought was, only cow butter. All the butter I eat was churned butter.

Were people good at paying the money back? They had to be. He wouldn't loan it to 'em if they didn't have some good security. He liked land. If they had some land, they got the money because they would have to put the land up for it. So they either paid the money back or lost their land.

Did anyone ever lose their land? Oh, yeah! [chuckling]

GEORGE MERTENS

Getting kids to school. In 1930, Dad signed a contract with the DeValls Bluff School District for a bus route, which included about six miles of dirt road (no gravel) and about nine miles of a state road which was gravel. That same year, Dad bought a 1930 Model-A truck and built a school-bus body to go on it along with screen windows and roll-up storm curtains. He was paid seventy-five dollars a month and had to furnish the bus, gas, tires, and maintenance. Tires, which were a big item back then, were very expensive and didn't last long. If Dad had to buy a tire, he lost money that month. Sometimes the dirt roads were so bad that he left the bus on the gravel road and transported the kids on the dirt roads with a four-up team of horses and a covered wagon.

At the height of the Depression, a lot of Model-T Fords were set up on blocks because people couldn't buy licenses for them. If you could sell your car, it would only bring about twenty-five to fifty dollars. We had a Model-T that was not licensed, so we used the school bus for transportation. The hired hand wanted to buy the Model-T so Dad let him work it out. When he got it worked out, he patched and aired up four tires and left one Sunday morning. That evening he came back on four rims. The car "set up" a long time after that. We had one hired hand that got married, moved into a house with his wife on the same salary, fifteen dollars a month.

EVELYN M. COONFIELD METCALF

Funeral preparations done in homes. In Vaughn, some of the people didn't have a lot of money, some zero. So when someone passed away, and the family didn't have the money to go to the funeral home, they did the next best thing. Today, a body is washed, cleaned up, and dressed. Back then, to preserve the body until the funeral, we used rubbing alcohol to keep the hands, face, and neck from turning black. These were the only parts of the body you could see. There are people who are still being buried this way.

Favorite Depression recipes:

Tomato Gravy: • 2 large ripe tomatoes, diced, or 1 cup diced canned tomatoes • 1/4 cup bacon drippings or cooking oil • 3 tablespoons white cornmeal • 1 1/2 cups water • Salt and pepper.

Directions: Whisk cornmeal, salt, and pepper (however much salt and pepper you like) into a large skillet with the oil, cooking and stirring over low heat until the cornmeal is a lovely golden brown. Turn heat to low, and stir in water, stirring all the time until the mixture thickens. Add the tomatoes and cook another 10–12 minutes or until hot.

Head Cheese • 1 large hog head • 1 1/2 tablespoons salt • Dash of red pepper • Dash of black pepper • 1 onion • 1 cup vinegar.

Directions: Clean hog head by removing eyes, ears, and brains. Saw into four pieces. Put in a large pot and boil until tender. Remove meat from broth. Pick out bones and cook onion until done in broth. Dip out onion and run meat and onion through food chopper Mix in peppers, vinegar, and salt, put in cheese cloth, hang, let drip overnight. Slice and enjoy. Refrigerate unused portion.

Lye Soap (boiled outdoors in a kettle) • 32 pounds of lard • 16 quarts soft water • 8 cans of lye.

Directions: Boil two hours and then add one more gallon of water. Stir and remove fire from kettle and pour into molds made out of clay. The tray was made out of a piece of old tin roofing, then nailed on four boards to form the tray. The soap was poured into the tray and let cool. In about one hour you could take a sharp knife, cut down the ridge in the tray then turn the tray over and tap on the top and the lye soap would fall out.

Then you would take the long roll of soap and cut it into bars—a longer bar for laundry soap, a shorter bar for handwashing and bathing. For gifts, Mother would mix in food color to give the soap a nicer smell and color. Then it was wrapped in a homemade tea towel or hand towel made out of feed sacks or sugar sacks as gifts when friends would come by.

MICHAEL FRANCIS O'CAIN

A love story. Rural southwestern Arkansas in the early 1900s was not a prosperous place for most of its citizens. These folks eked out a hand-to-mouth existence in the mostly forested hills and mountains taking on whatever presented itself in the way of daily work. Many worked in the timber industry. One of those was a man [named] Michael Francis O'Cain, my father-in-law, who was born in 1892.

Mike had occasion to be in the small town of Nashville one day in about 1914, to gather supplies for the logging operation for which he worked. As he came into town in his wagon, he passed some folks walking on the side of the road. One of them was a beautiful young girl. Mike, who was twenty-two, thought the girl who had caught his eye was probably in her late teens (turns out she was sixteen). They just locked eyes that day in 1914—not saying a word—as the wagon lumbered past. Well, he couldn't get her out of his mind, so he asked around the next time he was in town if anyone might know her name. Little did he know that she was making similar inquiries about the handsome young man on the wagon.

After learning each other's identities, they started writing letters to each other. After many months of letter correspondence, Mike asked Ruth Page to marry him. She accepted by letter and they set a date. By the time the date arrived, neither could clearly remember what the other looked like. But apparently it worked out because the marriage lasted until Mike's death in 1969.

The years between their marriage and his death were not easy ones. Unfortunately, they were astride that dark chapter of Americana known as the Great Depression (1929–1938). Mike had a number of different jobs during that period but mostly he continued to work in the timber industry and later for the Cotton Belt Railroad. Times were hard. It was a period when pennies were husbanded, as are ten-dollar and twenty-dollar bills today.

To feed his growing family, Mike decided to purchase a mule so that he could plant a large vegetable garden. He found such a mule nearby and agreed with the seller to a price of two dollars—twenty-five cents down and twenty-five cents each month until it was paid off. He took possession of the mule and began his gardening project with all of the eagerness he could muster. But, alas, times took a turn for the worst and he could not keep up the payments—even on those meager terms. He had to give the mule back to the original owner.

[Author's note: As told by his son-in-law, John E. McCown Sr., LTC, US Army (Ret.).]

LLOYD A. PERRY

Keeping food from spoiling before refrigeration. The best method we ever had was a smokehouse set on the bare ground, which helped control the temperature. It had double walls filled with sawdust, and there was also thick sawdust in the loft. It kept the cured meat at a good temperature so it didn't get rancid before we could use it in warm weather, and our home-canned foods never froze in winter.

We were always trying to protect our stuff from wild animals, insects, and rodents. Under our corncribs, we'd build up rock columns high enough we hoped the rats couldn't jump up to the logs, then turn old tubs and big pans over them before setting the smokehouse timbers in place.

Sometimes when I get to thinking back about those times, I remember the silly little songs we'd sing to amuse [ourselves] and pass the worrisome times. There was one I recall entitled "My Scattered Sweetheart." It was about her teeth being in one bowl, her eyes in another, her hair hanging on the back of a chair—it went on and on, with each verse telling about another part of her scattered around someplace, then ending with *"She's my sweetheart and I love her though she's scattered everywhere!"* Just silly stuff we could makeup or memorize and sing for fun. I guess because there wasn't much that was real that was funny. We had to get through the bad times somehow, and the reality was: It wasn't funny.

PAUL ROBERSON

Daddy whipped me five times, and I know where, why, and everything about that and the explanations he gave me. But one day I was riding the wagon from the area where we were working and gathering corn. Daddy sent us around by the store out at Boswell to pick up something. He had his glasses that he used when he was sharpening hoes, things like that. He'd left them in the box where we kept our lunch. I took them out and put them on my face and was messing around like I shouldn't have been. Somehow, those glasses got lost between when we left the field and we got back home. Well, I messed up because I got his glasses and he couldn't see to read the *Arkansas Democrat* and he wore me out. I remember that.

Another whipping that I got because of my cockiness when I was a kid. My daddy never laid anything by. We were getting the grass out of the cotton. The four boys and our dad would take five rows at a time, ten rows of a round. One time, we got down to the far end. I was tired and wanted to go to the house so I threw the hoe over my shoulder and said, "I'll see you back at the house."

"Paul, get back here and carry your row," my dad ordered. I just kept walking. Finally, I heard something go *"whack!"* He was cutting down an old horse weed—when they got full grown they were hard. He cut that horse weed down. When I heard that *whack!* I said to myself that I had better be getting back there or he'll wear me out.

I did. He wore me out anyway. But I needed all the five whippings I got when I disobeyed. My mother was hacking and slashing at me all the time.

We walked about a mile and a half through the woods to the church. We had to climb out of gully washes. My brother had me so scared of snakes that when we were walking through the woods, he would jump and holler, *"Snaaake!"* and I would take off. Then I would go home and tell Mother, and he'd get his come-uppance, too.

One time when I disobeyed my mother, I wanted to go swimming in the stock pond. So I got in it and Mother told me to get out, but I wouldn't get out. She came out to get me and I splashed water on her. That wasn't very smart of me. She got a little switch and worked on me. Since I was wearing Mother Nature's bathing suit, she could see the stripes on me where she had hit me and it broke her heart. She started crying and I started crying. But I never did that again.

Cardboard caskets for the poor. Back in those days there were people who were really poor. They couldn't afford to buy a wooden casket, but they would buy a reinforced cardboard coffin. Most of the time, there weren't many relatives of the deceased because if they lived too far away there was no way of getting there. Some of the preachers weren't ministers but just the laymen of the area who knew the Bible and some of the Scriptures like the 23rd Psalm or something like that.

Murray Funeral home would pick up the body and do the embalming.

The men in the Bethel community about three miles east of Okolona would always dig the graves and then have the funeral services in a little back-in-the-backwoods country church.

When my grandfather passed away, my mother borrowed me a pair of striped overalls from one of my cousins. That's how bad it was—we didn't even have decent clothes to wear. I had to borrow a sweater from a neighbor. We were not the only family that had that problem. When my daddy would buy me a pair of shoes, he would take a shoebox top and let me stand on it and he would trace my foot and bring it to JC Penney in Arkadelphia and buy me a pair of shoes.

PHILLIP WAYNE ROWAN

We traveled once a year to Hot Springs by wagon to market their produce. My grandfather, Phillip Wayne Rowan, was a tenant cotton farmer near Amity in Clark County, Arkansas, from the early 1920s until his death in 1932. He rented forty acres near the Caddo River bottom near present-day Highway 182 and was, by all accounts, a successful farmer.

Once a year, he loaded the wagon with what the family could spare in the way of vegetables as well as chickens, eggs, and the pelts of animals which he had trapped during the year. The family would set out for Hot Springs after lunch, taking with them bedding, a black cook pot and food to eat on the trip. When they reached the way station between Bonnerdale and Pearcy, they would stop for the night. For a small fee, the family could sleep in a large barn-like structure that had straw mattresses as well as a stable and feed for their animals. My grandmother would cook over an open fire in the black cook pot.

Early the next morning, they would drive on in to Hot Springs to the farmers' market. There, my grandfather would sell what he had on the wagon. This, along with the money from his cotton crop, was what the family had to live on during the coming year.

[Author's note: As told by Sammie Benjamin, Glenwood, Arkansas.]

UNION H. STOUDAMIRE

When my cousin wouldn't give me something to eat. My children's cousin came through the yard one day. He had been on this food program when it first opened up. I didn't know anything about it, probably because my children were so little I didn't get the chance to go to meetings and things where they told us what was going on. Anyway, my cousin come through my yard and went through the field to his home. He had two big slabs of bacon on his arm. I asked him for a couple of slices for my kids. But he would not give me even one slice. I just cried 'cause we didn't have nothin' to eat!

All the old folks came here by horse and wagon from South Carolina. They came by horse and wagon. They traveled by day and camped beside the road at night. It took them six weeks to get from South Carolina to Arkansas. The old folks would get together, sit down, and talk about it. I remember hearing them talk about how they had come from South Carolina in all those wagons.

They would drive so far in a day and then in the evening, they would stop and park on the side of the highway. They would stay there all night. They had feed for the horses. Next morning at four o'clock they would start out again. When they got to Arkansas, they asked, "Where are we going to settle?" They settled on Princeton Pike, just six miles from Pine Bluff.

Other memories. We would go in old fields to get broom sage to make brooms to sweep our floors. Also we would pull grass after the frost fall to make bed ticks (mattresses for our beds). Those was the good old days. [she smiles wistfully]

FRED THOMPSON

My dad opens a dry-goods store. When I was about six years old, my dad decided he had enough teaching, driving a school bus in Cave Creek and running ino problems of cashing in his school warrant so he opened a dry-goods store. After hearing of a little store in Lurton that was for sale, he said he was going to get into the grocery business or some kind of business. So we moved to Lurton where he bought a small dry-goods store. It didn't have much in it. But it was pretty good for a little town like Lurton.

He sold clothes, shoes, and tobacco. Of course, he went into debt to buy the store. He was pretty conservative—he didn't waste nothing. People would come in there—they were all in hard shape there, too. After coming into the store, they would get some of their needs, then say, "Mr. Thompson, put this on your sleeve, and I'll pay you just as soon as I get my milk check." Well, heck, there wasn't no more than a dozen cows in the whole county!

Did they ever pay him? No. When we left there eight or ten years later, my dad knew he had a big bill that people would never pay. "To heck with it!" he said, wadded up the bill, threw it away, and we got out of there. That was in about 1936 or 1937.

JAMES A. THOMPSON

Riding the cotton sack. My grandma always helped pick cotton every year. Granddaddy always raised cotton because that was our money crop. She was out there picking cotton—I'd go with her, I was just a little tot, maybe five or six—and she had this big long cotton sack that she pulled behind her to put the cotton in. It was always fun for me to jump on that cotton sack and just let her pull me for a little while. She told me twice that I had better not get on that cotton sack any more or I was going to get a spankin'.

I thought, "Oh, now, Grandma ain't gonna spank me." So I tried it a third. I jumped up on that cotton sack. She didn't look back at me, didn't say a word, just kept on picking cotton and pulling that sack. When she got up to the end of that row, she just stopped, unhooked that sack, and headed toward a big pile of brush out there at the end of the row. "Well, now," I thought, "what's she gonna do? I guess she's wantin' to go to the bathroom." But boy, in a minute, here she come back and she had a big ol' switch. I got one of the worst spankin's I ever got. I'm eighty-one years old now and it still hurts!

Fish fries and noodle fishing. Once a month all the neighbors would get together and have a fish fry. The men would jump in the creek and noodle fish, which was about the only way you could catch fish there.

Noodle fishing explained. They had a seine. The men would go down the creek with that seine and then back up the creek and they would catch a lot of fish. "Noodling" would be reaching your hand back under the bank of the creek and grabbing a bass by hand, bring 'em out, and put them in the seine. Once in a great while one would grab a snake by the tail. Once in a while, one of the men would get bitten but they didn't hardly pay attention to it, probably because it was just a water snake.

My job was to carry the fish after they had been caught back up to the fish fry, which would be about two or three miles. The women would be down there waiting for us. They would have a big fire going and they would be making biscuits and gravy. There would be about two dozen of us. We didn't know what hushpuppies was. Mostly, they would make corn bread.

Waiting for the top apple to fall. My granddad had an apple tree in his yard. It was the only fruit tree he had. We lived about a half a mile from him. One year that apple tree had an apple, right in the top of it. I thought, "Oh, boy! That apple's mine and I'm gonna get it!" It got up to a pretty good size and I would walk up to my granddad's front porch and I would sit on that front porch and look up at that apple.

"As soon as that apple falls," I'd say, "I'm gettin' it." I went up here every day because I didn't want that apple to fall and me not be there. After thirty days or more, I went up there one morning, got up there on the porch, looked up in the tree, and the apple was gone. I thought "Oh, my! Something has happened to my apple!" I jumped off the porch, went out there, and looked under that tree everywhere!

Finally, my grandmother opened the back door and says, "James, just what are you looking for?"

"I'm lookin' for my apple!"

"Ohhhh, that apple that was at the top of that tree?" she asked.

"Yeah, that's the one I was wantin'!"

Grandmother said, "I'll tell you where it's at. It's in the house in an apple pie." She had found that apple before I got up there that morning, went in, and made a pie with it. Now don't you think that was a nice thing to do so that everybody could get a bite out of it?

"What do you think about that?" she asked. A big lump come up in my throat and I thought, "Oh, I can't even talk!" I just turned around and walked off, went to my house, and the tears was rolling.

EVA SMOKE WELLS

The day our house almost burned down. We were planting corn in the field where my youngest brother now has his house. When my dad got to the end of the row, he turned the mule and looked in the direction of our house and saw it was on fire. He left the mule and we both ran to the house. Because my great-grandfather had not been willing to buy material for a new roof, we almost lost the house.

My cousin John lived across the field (where I live now) and when he heard Dad yell, he came running to help. My mom drew buckets of water from the well and I carried them to my grandmother. She carried the water to someone else on the stairs and this person handed it out the dormer windows to Dad and my cousin John. It was a hard-fought battle but we won. I could never understand why my granny had to go sit down and have a crying spell.

My first banana. When my oldest brother was born (at home), the doctor gave me and my mother's young brother a half of a banana. My grandmother took us out in the yard and we sat on a big rock and ate them. I think that was my first time to eat a banana. I was five years old. I am glad I was the first child born to Mom and Dad. I can remember when they laughed and played with us. Dad even went to the Tarzan movies with us. The three younger children never knew their parents when they were young and they missed a lot. Good times and bad, I've lived them all, and I think I'm a better person at solving problems because I've lived through so many.

ARWILDA WHITESIDE

Steps to making a pan of corn bread. Plant and grow seed corn, shuck it and shell it, take it to the mill in bags to get it ground into cornmeal, bring it back home, and make a pan of corn bread. Recipe: Two cups of meal, full cup of flour, two teaspoons of baking powder, one teaspoon of baking soda, one teaspoon of salt. Two teaspoons of sugar. Melt the shortening, use only buttermilk, let the skillet get hot, spread the meal in the bottom of the hot skillet, get it kind of brown, beat the two eggs, pour it in with buttermilk, and stir the pan. Add eggs at the last.

Homemade syrup: Put sugar and water on stove, add chopped orange peels we had saved, use the syrup on Mama's cornmeal pancakes.

Nellie the horse. She was a black, beautiful mare who we would hook to a wagon with Tom the mule and start on our journey. We would get to a spot on the road and in the water or creek and she would just stop. She would not go any further. We (Papa) tried everything we could to get her to go. She would kick, rare up, go backward, and try to tear up the wagon. It would take about an hour to quiet her down. Trips took a long time. Sometimes we never made it. It's funny now, but it was hard. A wagon full of children screaming, crying, hot, cold, raining, Nellie would not go.

Using our new icebox. After the flood, we got an icebox. To get the ice, we had to go in town, [which was] forty-five minutes away in the heat or cold. We wrapped it in a paper or cotton sack to keep it from melting in the wagon. We made homemade ice cream and lemonade. Mother had a place to keep her butter and milk.

CHARLES WHITFIELD

How Charlie built his first home. It was Grace's daddy's land and he let me build a house on it up there. He told me I could have the land if he had some money later on. He let me build a house there and I was going to pay him for the land. He died. When they settled up the land, Grace got this fifty-five acres here and we built a pine-pole house.

The poles would notch down, you know, and I built me a barn. I didn't have no money to buy good lumber. I sealed it on the inside of the logs with No. 3 lumber. Then I bought some building paper—it was cheap—and papered over that lumber. On the outside, I filled in them cracks full of mud.

I had that lumber inside so I just packed that mud in there. It was a warm house. I did most of the work myself. I had a [friend] cut the windows and doors out for me. I'd swap work. I didn't have no money. I would work for him two days and he would work one day for me. He helped me out that way.

The mud chimney. You ought to have seen that mud chimney I built! You'd put up a frame. I think I used peeled pine poles and sticks going around it. You'd work grass and mud worked in together and make what they called a "cat" back then and wrap it around them sticks, you know. The mud would dry and that would keep your chimney from catchin' a fire, and you'd have a good fire in it inside.

FLOYE WINGFIELD

Brother lost hand in sawmill accident. One of my brothers worked with Daddy in a little ol' sawmill he had. He had a little accident and sawed his right hand off the day before he was eighteen. It was along about three o'clock in the afternoon. Daddy came by from the mill to the school to see if somebody knew how to drive a car. He wanted to make a tourniquet. He was afraid Houston would bleed to death before he could get nine miles to town from near the DeGray community.

They got Houston to town and Dr. Townsend took care of his arm. It was still hanging on by just a piece of skin and he cut it in two. He cleaned up the hand real good and put it in a box and wrapped it in a little towel. When Daddy came from the hospital that night, he brought that with him. We all gathered around to see it. It was about our bedtime.

When we looked at that hand, we knew it was a part of our brother that we was never going to see again. We all cried. It was a funeral. The next morning, Daddy took it to a strong spring we had and buried it by the spring with a great big rock for a marker. And to this day we all know where that part of Houston is buried. But if that were to happen today, [Houston's severed hand] would go into the medical trash can and nobody would ever know what happened to that part of you.

The neighbor who stole their corn. One night we heard something in the back yard and went back to see what it was. We saw a man going away from the house carrying a tow sack on his shoulder. Claude, my oldest brother, said that was somebody stealing corn out of the crib. He'd just go get the gun and shoot whoever it was. I told him he'd better not go get that gun because Daddy doesn't allow us to play with that gun. Wilson, another brother, he knew who it was—ol' Joe. Then Claude said that Joe might be thinking of making some cornmeal for his family. By that time, Joe was out of sight. But since he didn't have any animals, I'm sure it was for food. They were hungry.

We never, never were in those circumstances. We raised our food but they didn't own anything to even make a garden from. They had nothing. Those were people who had it hard. One of them finally was making liquor for a man in the community who had the most money. That's how he got his money and he was sure he wasn't doing anything that would get him caught. But he did get caught and had to go to prison. That man was not a criminal. He didn't belong in prison. But those were the times. He was trying to get a bite of food for his family. Those were heartbreaking times.

JACK WOOD

We did a lot of hunting and trapping. We had a lot of cottontail rabbits on the farm. My younger brother Tom and I built about ten to fifteen rabbit traps that we would bait with a little piece of sweet potato. We got all the rabbits we needed. In fact, I sold them twice a week to the the main cook who was dressed in white on the passenger train that passed by close to us. The train would slow down every morning and I would hand these rabbits up on a coat hanger or something and he would catch 'em and go on. He would give me ten cents apiece for them, which he would leave the next day. If he picked up five rabbits today, as the train slowed down, he would drop off fifty cents the next day in a tobacco pouch. He never forgot. He would serve the rabbits to passengers on the train that carried a lot of people.

Things we didn't have. Television, radio, computers, electricity, air conditioners, gas for heat, running water, indoor toilets. I think the thing we needed most would have been electricity. We used lamps that used oil that fed the wick with oil up to the globe that would shine when the top of the wick was lit. All our heating was done by wood burning in a big fireplace and a nice iron cookstove that you could cook all day long on it. If the things we didn't have were shut down in this world today we would all go mad. We were facing the absence of these things and it didn't seem to matter because we never got used to having them.

The thing I would like to remember most is my mother and father. The way they taught us to deal with the hardships in life. If we got out of line, if they asked us to do something we would jump to and get it done. If we failed to comply we would get the belt. We didn't get away with anything like they do now. Most parents today aren't enforcing the rules at home and it's really causing school teachers problems.

[Author's note: From interview conducted by Maggie King, Jack Wood's granddaughter.]

Chapter 31

Survival Stories Our Young People Need to Hear

DOROTHY COX COSTON ASHLEY
I would try to teach them: How to raise a garden. • How to survive and be happy with very little material possessions. • How to get by without electricity and all of the modern conveniences it has brought into our lives. • How to make-do with whatever you had. • How to have fun without spending a lot of money. • That even if they failed and fell flat on their faces, the sun will still come up in the east the next morning. • That they are not "out" in the game of life until they decide for themselves that they are out. • That there actually is a living, loving God of the universe who longs for a personal relationship with them.

FRED NORMAN BLANKENSHIP
Number 1, get a good education. Number 2, take a foreign language course, probably Spanish. I would tell young folks—they don't like to hear it—that we are just blowing smoke when we tell 'em how hard times were back then—that they would really have a hard time now going through what we did because timber, cattle, and land have now replaced cotton as the money crops today. Way back then you could have bought land for a dollar an acre.

NANCY DUFFY BLOUNT and SPENCER DUFFY
NANCY BLOUNT: *No, they could not survive.* Because of the way we grew up, we learned to make an honest dollar and to manage our money so we could take care of the necessities with what we had. We were able to be "creators" in that we had to think and make things happen. Now, so much is given to our children, unfortunately, that people are killing themselves with problems they are having with foreclosures, losing jobs, and what have you. That's because they have never had the opportunity, I guess, to see what [hard times are] like, and to be forced to finding alternatives to bad situations.

SPENCER DUFFY: *The reason I think young people could handle another Depression is that in everybody is the instinct to survive.* And of course, we have a lot of safeguards and systems that will prevent vast numbers of people in America from being without the basic necessities. If those systems fail, however, I think we would have a problem. But then the whole country would have the same problem, and that would be terrible.

[Author's note: Mrs. Blount and Mr. Duffy are two of Leona Stith Duffy's children.]

LOUISE J. BRANT

How life was so different from today, that values were instilled in us, character as well as work ethics. I started out as a teacher and taught four years before I went over to library work. We went first thing in morning, we read the Bible. The kids got quiet, they respected it and they were ready to start the day. I think that had an awful lot to do with how our society has gone awry to this day. That was taken out of the schools, the children didn't get it at home in a lot of cases, and its affected their lives. People think you're really dated when you say that but I saw it happen.

What we lost is as important as other aspects [of education] and should be taught today. Every once in a while there is a ray of hope that they are going to do something to get it reversed but so far it hasn't happened. And I think something is missing from our schools right there. Schools are not what they used to be, either. As they come out of the schools today, students don't have the education they used to have. And I think that this lack of values and ethics has something to do with that too. Society has changed. Children just have too much coming at them that they didn't have back in the days when we were growing up on a farm and lived in our little ol' communities.

R. L. "BILL" CARTER

I don't think they would believe it if you were to tell them how bad things were. In fact of the business, when my grandkids hear it they want to brush it off like it never happened. Sometimes I wish we would have another one to show these young people what it was like.

Some of the people look down their noses at poor folks. But some of the best people in the world, according to my notion, are poor folks. I don't owe nobody nothin' now. Everything I've got is paid for and I got money in the bank. But I don't like to blow my money. I like to get my money's worth when I spend it. I remember one time we was choppin' cotton for a feller and we was making ten cents an hour for a ten-hour day.

Appreciate the value of money. The store out there at Rich had an ice box and had some Coca-Colas. Didn't have no electricity but they had an icebox. Had an ice man going around in a truck. But anyway, some of 'em crawled over the fence and went over there and bought 'em a nickel Coca-Cola. Man,

I wanted a Coke so bad I could almost taste it. But I got to thinking about it. I'd have to chop this cotton thirty minutes. [chuckling] So I turned it down.

WILLIAM EDWARD DELAMAR

Well, most of the young people I find today are too wasteful. They need to learn how to save. If you don't learn to save, you'll be forever in a strain. You've got to learn to do without. In my thinking, if you really want to see a very important person, there he is. [pointing to himself] You are more important to your own self than to anybody. Now think about who you want to be. Think about what you think about yourself. You either care or you don't care about your life.

I've been to a horse race just once in my life. It was over in Hot Springs. I paid a dollar to get in there. And I paid two dollars on a horse. My horse lost. I haven't been back since.

GENEVA KING EMERSON

Learning to "make do" made us efficient and inventive. Moral standards were set by the teaching of the Scriptures. We were expected to memorize verses from our Sunday School lessons, then apply them by the way we treated others, and through honesty and industry, caring for and working with others. Our elders often reminded us of those standards, sometimes quoting the Scriptures or telling stories from them. I think all that laid the foundation for the greatest society ever known. With such guidance, our hardships all worked out for good eventually.

BEULAH LEE MCLEOD EVANS

We were fortunate. We owned dairy cows and our own small farm. We raised our own gardens and orchards and canned a lot of food and raised our own chickens and hogs. Everything we sold didn't bring much, but we learned to get by. We made our own clothes—mostly from feed sacks. We had horses and mules, so we didn't burn gas. We made cross-cut ties and sold them to the railroads. We ate what we raised. My brother hunted and sold furs. We used wood in the cookstove and heaters. We cut and sold cedar posts. We were a close-knit family and we had good neighbors. Everyone helped each other.

RALPH EMERSON HALL

We didn't know anything different and trusted in God and our friends to get through the Depression. I would give anything to go back to those good ol' days. I would also tell them to concentrate on family and not so much on material things. I am still in touch with childhood friends and they all agree that our whole hometown was one close-knit family because we cared for each other and helped each other and were not selfish.

VIOLET HENSLEY

To me, [the Depression] was more like fun. I enjoyed playing on grapevines. Or feeding the hogs. Enjoy life as you go along. You're not gonna live but once. The way we lived and survived, there was another little money crop—we set traps for skunks and 'possums and sold the hide. We only got ten cents, but ten cents then meant more like ten dollars now. My first memory of buying a bar of candy was that it cost three cents. Now, it's more like a dollar.

I liked to get out and ride the horses. My sister and I would ride the horses from the field back to the farm. One time, we were playing and we had a mare—Cricket was the mule; she was mine and Belle was Edna's. Cricket wouldn't let the skunk ride with me. We killed rattlesnakes—I've got rattlers in my fiddle.

If you were asked by some young people today to talk about what it was like during the Great Depression, what would you tell them? It's like I said. I didn't realize I was poor but we were poor enough that if the flies had anything to eat, they'd have to bring it themselves. We saved all of our seed, we wouldn't throw nothing away, we cut the okra to the last bit we could get out of a pod of okra, we didn't waste anything.

I have a grandson-in-law that made himself a sandwich out of the wrong mayonnaise and he threw it out to the dog because he didn't like that mayonnaise. If he had been hungry enough, he would have ate that anyhow and would have been glad to have it.

MARY FRANCES LOVELL IZARD

The Depression taught us that money was not necessary for happiness. A little more would have helped but you take what you have and make the best of it. We had almost none of the things people today think they can't live without and we were happy. We had parents who loved us and we never went hungry. It wasn't a hamburger or a pizza but we have found out today that what we had to eat was a lot healthier for us. Things were not so rushed and there was time for family and friends. When you went to visit you sat and talked to each other instead of everyone staring at a TV. Creature comforts were few but we survived without them and we could do the same again if it became necessary.

ROSCOE E. JEFFERSON
When hard times come, you've just got to put up with it and not spend much money. Of course, you hear 'em talk on the news today about using credit cards. I never owned one in my life and I've gotten by. We growed what we eat. Times were pretty hard. If you had to walk everywhere you went now, you'd take it pretty hard. But that's what we done. We walked everywhere we went.

KENNETH GUY LACY
When telling my grandchildren of my experiences as a child, they look as though they can't comprehend. In reality, they can't. People would just have to experience those days to fully grasp the seriousness of the survival. I hope and pray our nation never will go through another time such as that. The last eight years of Republican government have pushed our economy to just about as close to the Great Depression days as can be pushed without going under. I am looking forward to our new president getting some work programs started similar to the way FDR did in the New Deal, and getting our nation's economy and spirit built back up so we can be the nation our forefathers envisioned.

EVELYN LANGLEY
Well, I'd like for them to know to have patience. We didn't have the stuff they have nowadays. We just had patience with what we did have, because that's all we knew. The big ones would help the little ones going back and forth to school, and things like that. Carry their books or their lunch. Lunch at school was carried in a tin bucket. And for us three girls, lunch was all packed together. There'd be a little jar of sorghum molasses and Mama would make some extra biscuits when she cooked breakfast and would put some biscuits in there. That's what we had for lunch.

WILLIS MAGBY
Young people wouldn't know how to handle a new Great Depression. They would go berserk. They would steal and rob. Back at the start of the Depression in 1929, people would ride freight cars because they didn't have the money to buy a ticket from here to Hot Springs. Sometimes people from around here would walk to Hot Springs or to Arkadelphia. Then they would catch a freight train and go from there to south Texas. They knew the ropes.

EVELYN M. COONFIELD METCALF

Children now have everything they want and don't need. We are too busy doing for ourselves today. We have no time for others, no concerns, no caring as people did back then. We didn't have much of the worldly goods [during the Depression] but more important, we had each other.

We don't even cook and prepare now like we did back then. We were always working in the garden starting in March, and then pulling weeds, hoeing, and sometimes replanting if needed. That took a lot of work. Then when the garden started producing, preparing them for canning took several days; picking, cleaning, cutting, and getting the produce ready for the jars. The jars had to be washed and sterilized with boiling water. The days were very full with canning, besides cooking, cleaning house, and doing the laundry by hand. We had to heat the water outside in a large pot over a fire pit.

The fields had to be planted and hay cut and pitched up into the loft for the animals. We were busy most of the time with work just to survive and sometimes not quite enough money to make ends meet. We had no electricity, of course. We had an icebox with a compartment to put a block of ice in. We bought the ice for a half cent and one cent per pound. We used the well to keep everything cool.

The young people of today wouldn't know where to begin to do what we had to do to live. It wasn't easy.

WILLIE MORRIS

What children are being taught today is against what this country is based on. Before that, the black children went to the black schools and the white children went to the white schools. After segregation ended, then everybody had to go to the same school. Before this happened, if a kid did something wrong, the principal could spank him. But after they got in the same school, the spanking stopped because black parents didn't want a white principal spanking their kids and white parents felt the same way about blacks whipping their kids.

Nowadays, I tell the kids that they are dressing themselves in ways that are so different than things were back when I was in school. Young girls are going around in what you call "low-rider jeans" and hip-huggers. And the boys walk along sideways and have to hold their britches up on 'em.

How well would young people today handle a new Great Depression? No, no! They haven't been taught at home. In other words, *"Train a child up while he's young in the ways he should go."* Nowadays, most parents who have children—both parents are working like the devil. They get up in the morning and take that baby to the babysitter. Come in in the evening and feed 'em and put 'em to bed. So many children!

WILLIAM PIERCY

The only thing I worry about, even about our own children and grandchildren, is that they don't know how to do without. Young people have everything they want at their fingertips. Young girls don't know how to cook, how to do washing, they can't sew (we learned to do that when we were just kids). They couldn't survive another Depression because of these reasons. I think there would be a lot of suicides. Many would turn to theft, stealing from neighbors and friends.

ALMA POUNDS

During the Depression, we just lived day by day and didn't worry about tomorrow. I hate to think of what kids would do today. Back then, most of us could eke out a living off the land someway or another. We knew how to use herbs and plants. You know, people do survive when they are out on their own and that's just about like what it was for us. You just have to survive. We knew what to eat and how to get it, how to use what we got. These kids today don't know how to get! I just don't think they would ever make it.

If you were asked, "Grandma, what was the Great Depression like?" what would you say? I'd have to say there was not much difference from not having a Depression because when you were poor as we were, we lived the same way as poor people did before the Depression. We didn't have to go barefooted in the winter or in the summer either. But there were kids who had to.

I became a teacher at eighteen. One winter morning it was snowing and the kids came to school. There were two boys, big ol' strapping fat boys, came to school without shoes on. I just couldn't stand it. They were in my class. I was the primary teacher in a four-room school. So the superintendent turned out school that day so the kids could go home. I'm sure those kids went home barefooted just as they did coming to school.

UNION H. STOUDAMIRE

I would tell the young people to go to school and get their education. Children was loving [during the Depression]. Everybody loved each other. I could write a whole notebook and still would not get all the stories told. But I am ending it for this time. I thank God that times are different now than it was then in the Depression days.

Could young people today survive a Great Depression? They might have been better off then than they are now. They wouldn't have money to get stuff that makes them crazy. I don't know.

Are things better or worse now than they were during the Depression? In a way they are worse today. Back then they didn't know anything about drug problems, things like that. Folks went to church. Back along then, boys would come by the house and take girls to church. Nowadays, they don't think about no church. We had stronger, more loving families back along then. They hadn't started gettin' wild like some kids are today. It's the older ones now who are wild or crazy or something!

JAMES A. THOMPSON

I'd say that neighbors would come to my granddad's place over on Cave Creek to get a little something to eat. They would come and ask my grandmother, "Do you have just a half a dozen eggs you let me have, or just a cup of flour, or something?" That's just how bad it was when you had to go to your neighbors and beg for something to eat. For us, we raised what we needed to eat. Except for once a year we would raise a hog. When we got ready to butcher that hog, all the neighbors would come in to help butcher it. When they got through, each neighbor would take some of it home with them.

EVA SMOKE WELLS

Our bank was a bucket. There were no real close neighbors. So we had to depend on ourselves if we got anywhere. I mean, we couldn't go next door and borrow something. "Next door" was three miles up the road! If you didn't have it, you did without.

To this day, I don't like debt. My daddy was the same way. Before he died, he had to go to the hospital. He kept telling me, "Get that money and keep it safe! I knew what he was talking about because up in the closet of his bedroom he had coffee cans stuffed with money. We took him to the hospital about three o'clock in the morning. He was hurting so bad but there as nothing we could do. Actually, it turned out to be diverticulitis, so there was a lot of pain to it. So I left him down there with my brothers and I came back and got the money. I put it in a big, white bucket and hung that bucket from the ceiling of my shed out here, just like there was nothing valuable in it. It stayed there as long as he was in the hospital. Mother was in there, too, and they finally wound up in the same room.

She got out one day and he got out the next. That night, I had to stay over there with her so I came and got the bucket, put the money in it, and took it over there and set it by the fireplace and that's where it stayed that night. After everybody else had gone, I said, "Daddy, you want to count this money because it's going in the bank." Turns out he had nine thousand dollars in that closet!

To open an account, the bank said they needed two signatures. Mother had lost a leg because she was a diabetic. There was no way she could get to the bank so I was one of the two co-signers with Daddy. Mother didn't like that a bit. I said to her, "Mother, I don't need your money. If something happened to Daddy, I could get the money for you. But you can't go down there."

One time I finally got her into a car, took her to town and to the bank, and then I just drove her around to places she'd never been before in the area. And that's the last time she was out in a car. She died about a month later.

ARWILDA WHITESIDE

Be thankful that they don't have to do what we did during the Depression to get their food. If they did, they would probably die! But we raised everything. We didn't know we were poor. We just did the best we could and stayed happy. We were very thankful for everything our parents taught us: Never to touch anything that didn't belong to us, be honest, be respectful, most of all, to fear God. He never sleeps. He sees everything we do, and there is a Heaven and a Hell.

Do you think young people today would be able to handle something like the Great Depression? No. One thing, and I have to plead guilty, we have pampered them too much. We say, "We had it so hard and we don't want you to have it harder than we did." Our kids know how to cook. We weren't poor and we had fun cooking because it taught them how to take nothing and make something out of it. We would cook, we would make things. In fact, we made our own Christmas gifts and stuff like that.

We had fun but a majority of the kids now, all they can think about is going to a McDonald's and places like that. But today, not only the young people but the mothers don't know how to make a biscuit. They have to go and get some Jiffy mix. Or how to make a pan of corn bread, that kind of thing. So another Depression is going to be very hard, not just on young people but on older people, too.

My husband is eighty-seven years old. He gets up each day and goes out and works on our garden. He cuts the grass. Our grandkids tell him, "We'll be over there to help after a while," but by the time they get here, he's finished. When I was young and living at home with my mama, I said there were three things I wasn't going to do when I grew up. One was I wasn't going to get up early; two, I wasn't going to take any cod-liver oil and no castor oil; and three, I wasn't going to pray because we couldn't do nothing unless we prayed. If somebody came over to the house, we couldn't eat until we prayed. But today I pray for everything!

My daddy would say, "We've got to pray, we've got too pray!" One day we were playing and—kids will be kids—didn't pray. We had a chinaberry tree outside the yard and one of the kids threw a whole chinaberry into another child's eye. It got up under his eyelid. And you talk about praying! "Oh, God!" I prayed. "How are we gonna get that chinaberry out?" I don't know to this day how we got that chinaberry out, but it came out.

Did you pray about it? Did I pray about it? We ALL prayed about it! We weren't praying. We were screaming! We knew we were gonna get a killin' because we got a whippin' for everything!

What would you tell young people today is the secret of happiness? Fear God and love Him. We have gone through many, many, many crises. We're still going through crises. [patting the Bible] That's the secret. That's the secret. We thank God for taking two little nobodies—that's what I call us—and using us. Sometimes I wonder. I say, "God? Is it us you want to use in so many different projects?"

We had twelve children—two of them dropped out of high school but they went back. And they made it. God used us and He is still using us. We're working on a project right now, "Make a Difference," a program for all kids who are in trouble—black or white. They will be welcome. We want to turn Pine Bluff into a city of love.

> [Author's note: Marquette Whiteside, one of their children, was one of three American soldiers on the cover of *Time* magazine (December 29, 2003–January 5, 2004), "Person of the Year" issue, for having captured Iraqi leader Saddam Hussein.]

CHARLES WHITFIELD

I don't know if you could get young people today to believe what you was telling them. You'd tell 'em all these things you done and they would say, "Oh, he's makin' that up! They'd make fun of you if you told 'em you built a dirt chimney but that's mostly the way they were built back then. Very few people in the country had a brick chimney. I'd tell 'em to stay off liquor and dope. Never be a crook. Pay your debts, if you can. If you can't, don't deny them.

ALBERT M. WILLIAMS

They tried to teach us kids what was right, how we were supposed to act. And they insisted that we get that done. Education? They encouraged us up to a point. I guess that at that time they didn't know any better about going on beyond a certain point. But they believed in schooling and tried to impress on us its importance. They didn't get much further than that because they didn't know any better. They did the best they could. That's all I can say. I was the very first member of the family to go to college.

What would you tell young people about what it took to get through the Great Depression? Hope and courage. They must have these values to do it. "Do the best you can with what you have where you are" is something I heard many times. I sure have.

LUCILLE RIDER WILSON

It was strengthening. You learned that you could depend on yourself. You learned that you could do anything. Whatever needs to be done, you can do it. You've met every demand that was made on you. It teaches you not only how to survive but to face whatever comes. To this day, if you ask me to do something, I will agree to do it because I know I can do it. My sister and I have talked about this. That was one of the positive things for me that came out of the Great Depression. We didn't know we could fail.

Daddy was tall, thin with narrow shoulders. He moved quickly and purposefully. He told everyone in the family what to do and how to do it! He always did it in such a manner as to make us want to please him. We were a strong family unit because of his leadership. My father, who barely was a reader, believed in education. I got my education because my daddy believed in it. When he picked me off the farm and put me in college, he made my life what it is today. He opened up the world for me. Daddy is my hero.

FLOYE WINGFIELD

I would tell them that I had had more than my share of blessings, starting with the Great Depression. The Lord has let me live and understand what the Great Depression was. But we accepted the blessings as hardships. We didn't have drugs. But today we have enabled children to do nothing but look for entertainment. The Depression also taught me how to be truly grateful and thankful for anything we have, to stop and give a little thought to where it came from. We came into the world with nothing and we will leave with nothing.

So let's thank our Father in heaven for what he has allowed us to have while we were here on earth.

There was a lot of blessings during that time. We need to look for some of those blessings and not let history repeat itself because it will. You look at our stock market, our money situation, how fast we are losing the middle class. We're fast dividing into the have's and the have-nots. Let the Depression teach them that they need to do something besides play now. There's more to life than a video game.

How did you come up with the will to keep on going on? Well, you didn't have anywhere else to go, did we?

IV. Epilogue

As this three-year project has ended, I know that there are at least hundreds of Depression-era farm families that I have missed. I also realize that there are those of you listed among the survivors who were quoted more or less than others or not quoted at all. For those of you in the latter category, please know that your memories are an integral part of this book but that I was faced with space limitations.

I will never forget the survivors or their stories. Some are sad, some are funny, all are inspiring. Here is one last response: Geneva King Emerson's childhood memory of the courage and faith of Depression-era farmers:

"I'll always remember riding in the wagon bed with my siblings behind my parents or grandparents up front in the wagon spring-seat, and thinking about that box of cakes or pies and fried chicken to be spread at the church dinner. Or riding along at night as we went to a revival meeting and watched the beautiful full moon. When I was very young, I thought the moon followed us! Sometimes we sang the gospel songs as we rode along. Then there were the nights after the day's work in the fields, after chores and supper, when my parents and young aunts and uncles gathered in on my grandparents' porch with the lamp extinguished—kerosene cost about a dime a gallon and we had to make it last a long time.

"There were few radios and no one had dreamed of a TV. In my mind, I can still see the big moon rise over the hill as the family rested and visited. We sang the same old songs of faith, which seemed to be more healing than medicine, listened to the 'katydids' or crickets later in the season, the screech owls and hoot owls. A little child—a very blessed little child. I soaked it all in; dreamed on it, grew on it. When I come to the final sleep, I think that will be my last memory."

I am deeply grateful to each of you for providing me with one of the greatest experiences I could ask for in this life—telling your stories of survival. To each of you, a heartfelt "Thank you!"

William D. Downs Jr.
November 25, 2010

V. Appendix

Stories of Survival: Arkansas Farmers during the Great Depression

A project funded by the Arkansas Humanities Council • William D. Downs Jr., Ph.D., Project Director • P.O. Box 3791 • Ouachita Baptist University • Arkadelphia, AR 71998 • 870/246-5390

Date of interview:

___ Interview release on file

Family information:

*Name: _____ * Date of birth: Your age:

*Place of birth: _____

*Present address: _____ City/State/ZIP _____ County _____

Telephone: ___ / _____ Cell ___ / _____ e-mail: _____

* Spouse's name   Please check one: ___ *Living* / ___ *Deceased*

Names of your children:

Your parents' names:

* Contact person *(if needed)*: _____ Relation *(son, daughter, etc.)*:

Contact telephone: ___ / _____ Cell: ___ / _____ e-mail: _____

1) How old were you during the depression years? (1929–39)

* Earliest memories of the Great Depression:

2) How many persons were living in your home? What were their relationships *(number of brothers and sisters)*?

3) What was the size of your farm?

4) Did you own your own farm, or did you rent the land? Explain as needed.

5) If you rented the land, were you a sharecropper?

6) What kind of crops did you raise?
   • How did you work your crops?

★ 7) Did you know you were poor?

8) What chores were assigned to family members?
   • How did you keep your milk from souring?

★ 9) To what extent did the weather—droughts, floods, etc.—affect your crops?

10) How did your family make enough money to pay for seed, etc.?

11) What was the source of your drinking water?
    • What did your family do about bathing?
    • How often did you bathe?
    • How did you wash your clothes?

12) What kind of transportation did you have?
    • How long did it take to get to the nearest town?

★ 13) What is the highest grade level you achieved in school?
    • In as much detail as possible, describe your school experience *(how you got to school, description of the schoolhouse, games played, etc.)*.

★ 14) What about medical care during this period? Home remedies?

15) How did you get your news?
    • From neighbors? Newspapers? Radio?
    • If radio, what were your favorite radio programs?

16) What about clothing, shoes, etc. Were they purchased or homemade?
    • Did you ever go barefooted? *(Share memories)*

★17) What were the most serious problems your family faced as farmers during the Depression? *(Recall such experiences as lack of money, bank closures, loss of credit, sickness, floods, droughts, etc)*.

18) How did you celebrate holidays *(Christmas, Thanksgiving, birthdays, etc.)*?

★ 19) What role, if any, did your spiritual life, neighbors, etc., play in how your family endured the hard times during the Great Depression.

20) As a child, how did you and your friends entertain yourselves *(marbles, games, etc.)*?

21) What about dating? Where did you meet your spouse?

★ 22) What "stories of survival" from the Great Depression would you like to pass on to our young people today?

★ 23) What is your *worst* memory of the Great Depression?

★ 24) What is your *best* memory of the Great Depression?

★ 25) What is your funniest memory of the Great Depression?

★ 26) Other: *(What I didn't ask about, memories, favorite stories, little incidents that didn't seem to be important at the time but have remained in your memory, etc.)*

Questionnaire revised 8-11-20

Interview-Subject Release Form

"Stories of Survival: Arkansas Farmers During the Great Depression"
William D. Downs Jr., Ph.D., Project Director
2501 Wilshire Drive • Arkadelphia, AR 71923
Home: 870/246-5390 • Cell: 870/245-6312 • e-mail: downsw@sbcglobal.net

Because I have special information about a significant event in American history, the Great Depression, I agree to be interviewed by William D. Downs, Jr., for the oral history project, *"Stories of Survival: Arkansas Farmers During the Great Depression."*

I am aware that interview recordings and any transcripts or other versions of the interviews that might be created may be made available for historical and other research and for public presentation by such means as media productions, museum exhibits, and web site presentations, and for use in educational materials for public and private schools, colleges, universities. Participation in this oral history research project is entirely voluntary, and I understand I may withdraw from the project at any time.

I have read the above and give my consent for recordings of interviews with me and any possible transcripts or other representations of those interviews—both written and photographed—to become part of the archives of the Arkansas Humanities Council, Ouachita Baptist University, libraries and/or other archival collections I designate, where they will be made available for historical and other research and for public presentation. I hereby assign rights, title, and interest, including copyright, pertaining to interviews with me to the interviewer named above or I designate.

_____

Interviewee (*Please print*)   Interviewee (*Signature*)

_____

Interviewee address   City   State   ZIP

Interviewee telephone: _____ / _____-_____ E-mail_____

Date signed:_____

Notes

1. David M. Kennedy, *Freedom from Fear: The American People in Depression and War, 1929–1945* (New York: Oxford University Press, 1999), 208.
2. Vernon Massey, "The Great Depression in Augusta," *Woodruff County Historical Society Quarterly* 11, no. 2 (spring 1983): 5.
3. Nancy Weaver Williams, "The Great Depression: A Terror or a Blessing?" Paper written for Principles of Macroeconomics class, University of Arkansas, Fayetteville, fall 1999.
4. B. C. Hall, "Life Before and During the Great Depression in the Arkansas Ozarks," *Carroll County Historical Quarterly* 25, no. 2 (summer 1980).
5. Nancy Weaver Williams. Unpublished documents.
6. Interview with Wiliam Piercy, Survivor, February 5, 2009.
7. Ben F. Johnson III, *Arkansas in Modern America, 1930–1999* (Fayetteville: University of Arkansas Press, 2000), 2.
8. John M. Barry, *Rising Tide: The Great Mississippi Flood of 1927 and How It Changed America* (New York: Simon & Schuster, 1997), 174.
9. Barry, *Rising Tide,* 188.
10. Johnson, *Arkansas in Modern America,* 1.
11. www.loc.gov: The Great Depression and World War: 1929–1945.
12. *Southern Standard,* January 8, 1931.
13. *Southern Standard,* December 25, 1930.
14. Donald A. Holley, "Arkansas in the Great Depression," *Historical Report of the Secretary of State, Arkansas,* 160.
15. Gail S. Murray, "Forty Years Ago: The Great Depression Comes to Arkansas," *Arkansas Historical Quarterly* 29, no. 4 (winter 1970): 299.
16. Johnson, *Arkansas in Modern America,* 10.
17. Nancy Weaver Williams. Unpublished document.
18. Massey, "The Great Depression in Augusta," 5.
19. Holley, "Arkansas in the Great Depression," 159.
20. "Arkansas Fights for Its Life," *Literary Digest,* February 28, 1931.
21. Johnson, *Arkansas in Modern America,* 11.

22. Charles A. Walls, "Eyewitness Account of the England Food Riot," Handwritten manuscript published in *Lonoke Democrat,* April 16, 1931.
23. Lement Harris, "An Arkansas Farmer Speaks," *New Republic* 67 (May 27, 1931): 40–41.
24. Johnson, *Arkansas in Modern America,* 11.
25. Tom Dillard, "Drought of 1930 Left Some in State Desperately Hungry," *Arkansas Democrat-Gazette,* September 4, 2005, 5-H.
26. Murray, "Forty Years Ago," 298–99.
27. Holley, "Arkansas in the Great Depression," 158.
28. Jeannie M. Whayne, Thomas A. DeBlack, George Sabo III, and Morris S. Arnold, *Arkansas: A Narrative History* (Fayetteville: University of Arkansas Press, 2002), 320.
29. Massey, "The Great Depression in Augusta," 2.
30. *Yellville* (Arkansas) *Mountain Echo*, September 3, 1931.
31. *Yellville* (Arkansas) *Mountain Echo*, August 27, 1931.
32. Holley, "Arkansas in the Great Depression," 174.
33. www.The First 100 Days.huppi.com.
34. www.The First 100 Days.huppi.com.
35. Alex Daniels, "State's Robinson Sold New Deal in Congress," *Arkansas Democrat-Gazette,* December 26, 2008, 1-2A, 17-A.
36. Daniels, "State's Robinson Sold New Deal in Congress."
37. William J. Atto, "Brooks Hays and the New Deal," *Arkansas Historical Quarterly* 67, no. 2 (summer 2008): 171.
38. Kennedy, *Freedom from Fear,* 205.
39. www.u-s-historory.com/Agricultural Adjustment Act.
40. www.Encyclopedia of Arkansas/Arkansas Writers Project.
41. Daniels, "State's Robinson Sold New Deal in Congress."
42. Holley, "Arkansas in the Great Depression," 174.
43. Trey Berry, *The Arkansas Journey* (Salt Lake City: Gibbs Smith, 2007), 196–97.
44. Holley, "Arkansas in the Great Depression," 168.
45. Kennedy, *Freedom from Fear,* 209.
46. Doug Smith, "Sharecroppers on the March: The STFU comes alive in Tyronza," *Arkansas Times,* May 4, 2007 (Web page), 15.

47. Kevin Freking, *Arkansas Democrat-Gazette,* NW edition, April 29, 2001, A-1.
48. www.The Encyclopedia of Arkansas History and Culture/STFU.
49. Whayne et al., *An Arkansas Narrative,* 335.